Time Series Forecasting Using Generative AI

Leveraging AI for Precision Forecasting

Banglore Vijay Kumar Vishwas
Sri Ram Macharla

Apress®

Time Series Forecasting Using Generative AI: Leveraging AI for Precision Forecasting

Banglore Vijay Kumar Vishwas
San Diego, CA, USA

Sri Ram Macharla
Montville, NJ, USA

ISBN-13 (pbk): 979-8-8688-1275-0
https://doi.org/10.1007/979-8-8688-1276-7

ISBN-13 (electronic): 979-8-8688-1276-7

Managing Director, Apress Media LLC: Welmoed Spahr
Acquisition Editor: Celestin Suresh John
Editorial Assistant: Kripa Joseph

Cover designed by eStudioCalamar

Cover image designed by Freepik (www.freepik.com)

Distributed to the book trade worldwide by Springer Science+Business Media New York, 1 New York Plaza, Suite 4600, New York, NY 10004-1562, USA. Phone 1-800-SPRINGER, fax (201) 348-4505, e-mail orders-ny@springer-sbm.com, or visit www.springeronline.com. Apress Media, LLC is a California LLC and the sole member (owner) is Springer Science + Business Media Finance Inc (SSBM Finance Inc). SSBM Finance Inc is a **Delaware** corporation.

For information on translations, please e-mail booktranslations@springernature.com; for reprint, paperback, or audio rights, please e-mail bookpermissions@springernature.com.

Apress titles may be purchased in bulk for academic, corporate, or promotional use. eBook versions and licenses are also available for most titles. For more information, reference our Print and eBook Bulk Sales web page at http://www.apress.com/bulk-sales.

Any source code or other supplementary material referenced by the author in this book is available to readers on GitHub. For more detailed information, please visit https://www.apress.com/gp/services/source-code.

To my parents

Rathnamma *and* **Vijay Kumar**, *who nurtured the seeds of knowledge planted within me.*

—Vishwas

To my Mom and Dad

Mrs. Satyavathi Macharla, *Retd. Mgr ECIL*

Mr. Narayana Murthy Macharla, *Retd. Mgr ECIL*

And in memory of my grandfather **Mr. M.V.S.N. Murthy**. *100% of the royalty I receive from the sale of this book will be donated to* **St. Jude Children's Research Hospital**.

—Sri Ram Macharla

Table of Contents

About the Authors..ix

About the Technical Reviewer xi

Acknowledgments xiii

Introduction xv

Chapter 1: Time Series Meets Generative AI 1

What Sparked Interest in Time Series?...1

Introduction to Time Series Analysis..2

    1.1 Characteristics of Time Series Data 3

    1.2 Time Series Forecasting Methods 4

    1.3 Introduction to Generative AI 8

    1.4 Evolution from AI to Generative AI 9

    1.5 Generative AI with Time Series...13

    1.6 Introduction to Large Language Models 15

    1.7 Summary 16

    1.8 References 16

Chapter 2: Neural Networks for Time Series 17

2 Introduction to Perceptron 17

    2.1 Technical Overview of a Perceptron 19

    2.2 What Is Multilayer Perceptron? 21

    2.3 CNN-Based Architecture for Time Series..............................27

    2.5 Neural Networks for Sequential Data 50

2.6 Neural Networks Based on Autoregression 64

2.7 Neural Basis Expansion Analysis 74

2.8 Summary 80

2.9 References 80

Chapter 3: Transformers for Time Series 83

3 Introduction to Transformers 83

3.1 Technical Overview of Transformers 84

3.2 Vanilla Transformer 94

3.3 Inverted Transformers 102

3.4 DLinear 109

3.5 NLinear 118

3.6 Patch Time Series Transformer 122

3.7 Summary 129

3.8 References 130

Chapter 4: Time-LLM: Reprogramming Large Language Model 131

4 Fine-Tuning vs. Reprogramming 132

4.1 Technical Overview of Time-LLM 133

4.2 Time-LLM in Action 138

4.3 Summary 153

4.4 Reference 154

Chapter 5: Chronos: Pre-trained Probabilistic Time Series Model ...155

5 Introduction 155

5.1 Technical Overview of Chronos 156

5.2 Time Series Tokenization 157

5.3 Training 158

5.4 Inference 159

5.5 Chronos in Action...159

5.6 Summary 167

5.7 Reference 167

Chapter 6: TimeGPT: The First Foundation Model for Time Series....169

6 Introduction...169

6.1 Technical Overview of TimeGPT...171

6.2 TimeGPT in Action..173

6.3 Summary 182

6.4 References 182

Chapter 7: MOIRAI: A Time Series LLM for Universal Forecasting....183

7 Introduction...183

7.1 Challenges with Building a Universal Forecasting Model....................184

7.2 Technical Overview of MOIRAI 186

7.3 MOIRAI in Action 188

7.4 Summary 194

7.5 Reference 194

Chapter 8: TimesFM: Time Series Forecasting Using Decoder-Only Foundation Model..195

8 Introduction...195

8.1 Technical Overview of TimesFM 196

8.2 TimesFM in Action 199

8.3 Summary 209

8.4 Conclusion 209

8.5 Reference 210

Index..211

About the Authors

Banglore Vijay Kumar Vishwas is a seasoned Principal Data Scientist and AI researcher with over 11 years of experience in the IT industry. He is currently based in San Diego, California. Vishwas holds a Master of Technology in Software Engineering from Birla Institute of Technology and Science, Pilani, India. He specializes in developing innovative solutions for large enterprises, with expertise in machine learning, deep learning, time series forecasting, natural language processing, reinforcement learning, generative AI, and AI agents. He is the author of *Hands-on Time Series Analysis with Python* published by Apress. He is the inventor of a patented method that utilizes AI to minimize emissions from gas turbines.

Sri Ram Macharla is a consultant and architect in the areas of AI and ML with over 19 years of experience in IT. He holds an MTech from BITS Pilani and has experience working with clients in domains such as finance, retail, life sciences, defense, and manufacturing. Additionally, he has worked as a mentor, corporate trainer, and guest faculty teaching AI and ML. He has papers published and works as a reviewer with leading journals and publishers. He is passionate about mathematical modeling and applying AI for social good. He is currently affiliated with Involgixs Inc.

About the Technical Reviewer

 **Sai Chiligireddy** is an Engineering Manager at Amazon with a decade of experience in software engineering, specializing in generative AI, cloud, and distributed systems. Beyond his professional role, Sai is passionate about mentorship. He actively supports new engineering managers, senior engineers, and university students, mentoring them on career development and technical expertise.

Acknowledgments

This book would not have existed without the tenacious support of my incredible family. To my parents, **Vijay Kumar** and **Rathnamma**, whose love and guidance have been my guiding light. Thank you for your endless belief in me. Your sacrifices and constant support have paved the way for success in my life. To my wife, **Janani**, my rock and my biggest cheerleader, thank you for your unwavering love, constant encouragement, and indomitable support throughout this challenging journey. Thank you for your patience and understanding. To my brother, **Shreyas**, thank you for our unbreakable bond and the unflinching support that has always been there for me. And finally, to my son, **Hiyan**, the most amazing little human I know, may you always chase your dreams with boundless enthusiasm and perhaps one day write your own book.

—Vishwas

I would like to thank my spouse, **Meena**, and son, **Sudhish**, for taking up my responsibilities around the house while I was busy working on the book.

Writing a book of this sort is impossible without the motivation and support of friends and well wishers. Thank you Dr. Damahe, Raju Gandhi, Aaron Maxwell, and Ganesh Samarthyam; your articles and responses to my mails were motivating. To my former colleagues – Dr. Anji Pasala, Sridhar Murthy, G. Madhu, S. Karthikeyan, and Hari Sharma – for the opportunities, support, and guidance. To my friends – Focus group, Appalachari group, Sudhir Sriramoju, Irfan Chavda, Jaime, N. Uday, Naveed, Mallik Katta, Anuj Mohan, Naga Kishore, S. Koley, Chaitu Tanuku, Madhu Kanala, and Balu Nayak – always appreciate and thank you for the

ACKNOWLEDGMENTS

support. My sister and brother-in-law, whom I can always fall back on, thank you. Sharad Chilukuri, Director at Involgixs, for encouraging and supporting this initiative. My former and current supervisors – Gladson, Sandhya, Muthu, Odie, Martin, Tim M., Chandan, Ren, Manoj, Saurabh K., K.S.N. Murthy garu, and others – for providing me the opportunity to work on high-impact projects and for the guidance. To the organizations that provided me the opportunity to work on corporate training assignments – thank you for the trust in me. Thank you Dr. Sudhakar, Dr. V. Uday, and Dr. A.V. Ramana for always being around for any technical discussion. Dr. Nicoleta Serban, thank you for the amazing course on time series analysis. It helped in laying a strong foundation.

Lastly, I would like to thank my coauthor Vishwas for the numerous arguments and discussions to ensure we revise the content and do our best.

—Sriram

We would like to express our appreciation to Sudhish Macharla, Praveen Nandan K, and Siva Pichappan for their contributions in proofreading the early draft and testing the code.

We express our heartfelt gratitude to T. Sowmya for her invaluable assistance in answering all our questions throughout the development process. We would also like to thank Celestin John for his guidance in refining and approving the proposal. Finally, we extend our appreciation to all the reviewers and the entire production team at Apress for their contributions.

—Vishwas and Sriram

Introduction

"Guru Brahma, Guru Vishnu, Guru Devo Maheshwara, Guru Sakshat Parabrahma, Tasmai Shri Gurave Namah" – a disciple expressing gratitude and reverence toward their guru (teacher).

Grateful to my gurus who guided and supported me in the form of teachers and friends.

A couple of years back while working on a project related to time series, we wanted to explore newer techniques in forecasting to improve precision. The advent of GenAI provides us with an opportunity to explore LLM-based models for forecasting. However, there was not enough material to help the team come up to speed. The research papers were difficult to understand for the team who came from diverse levels of mathematical backgrounds, so we had to go through a steep learning curve.

We were looking for a resource that would equip us with the theoretical understanding of the models and practical implementation with python sample code. We could not find any, so that gave birth to the idea of writing this book. We present this book that is catered to the needs of working professionals to come up to speed. Those who wish to dive deeper may want to read the reference papers after reading this book.

This book is primarily targeted toward intermediate to advanced time series forecasting modelers. So if you are a beginner, we suggest you to pick up a beginner-friendly book like *Hands-on Time Series Analysis with Python* by Vishwas and Ashish before reading this book. Researchers are suggested to read the provided references after going through this book.

The book starts with a motivation to learn time series forecasting. Chapter 1 introduces different time series techniques, generative AI, large language models, evolution, and milestones to date.

Chapters 2 and 3 discuss neural networks and transformer theory and implementation. You can use these chapters to refresh your knowledge and learn to implement them by leveraging modern tools.

Chapters 4–8 cover topics related to foundation models for time series forecasting. Each chapter discusses a new foundation model. We begin by understanding the technical overview, relevant concepts, and implementation using Python code and libraries. Techniques that help to understand forecasting by repurposing and reusing foundation models meant for NLP are explained.

All chapters (except Chapter 1) discuss how to implement the models with a dataset and full code with explanation. Where possible and applicable, we try to implement the models for both univariate and multivariate scenarios.

CHAPTER 1

Time Series Meets Generative AI

Chapter Goal: Introduction to time series, evolution of artificial intelligence, and a gentle introduction to generative AI and large language models.

What Sparked Interest in Time Series?

There is a lot of buzz in the IT industry about NLP, computer vision, generative AI, transformers, and AI agents. However, a specific use case encountered while working on a consulting project for a manufacturing client, which was solved using time series techniques, captured interest in time series.

For over two decades, a team relied on a legacy approach using moving averages to forecast product demand for the next year. This system, however, often resulted in inaccurate forecasts, leading to significant waste due to under- or overestimation and instances where orders couldn't meet actual demand.

A more sophisticated approach was implemented using simple ARIMA (Autoregressive Integrated Moving Average) models to address this issue. This upgrade significantly reduced waste and, to our knowledge, has eliminated instances of underestimation since its implementation. While

© Banglore Vijay Kumar Vishwas and Sri Ram Macharla 2025
B. V. Vishwas and Sri Ram Macharla, *Time Series Forecasting Using Generative AI*,
https://doi.org/10.1007/979-8-8688-1276-7_1

this project was less complex than other initiatives using computer vision and NLP, the time series solution delivered immediate cost savings and empowered the team to make informed decisions on time. This success also garnered significant recognition from senior management.

Introduction to Time Series Analysis

Time series analysis is a statistical and advanced mathematical technique for analyzing time-dependent data. It is used in various fields such as finance, economics, healthcare, environmental monitoring, marketing and sales, energy and utilities, manufacturing, telecommunications, engineering, and many more to identify patterns within data over time.

The goal of time series analysis is to identify the underlying patterns, trends, and seasonality in the data and use this information for making informed predictions about future values. Let's put this in context through some real-world examples.

> **Example 1**: Predict inventory for supply chain optimization.
>
> **Example 2**: Predictive or preventive maintenance is a proactive way to maintain equipment health, machinery, or other assets in optimal condition to prevent breakdown.
>
> **Example 3**: Forecast pandemic spread.
>
> **Example 4**: Identify patterns for the bullwhip effect and cart loading (refer to the "Summary" section).

1.1 Characteristics of Time Series Data

a) **Time dependence**: Data points are ordered in time and have a natural temporal sequence, which means that prior observations frequently influence the value of each observation.

b) **Autocorrelation**: Statistical measure that describes the relationship between an observation in time series and its own past values.

c) **Stationarity**: Statistical properties of time series do not change over time.

d) **Nonstationarity**: Statistical properties, like mean and variance, change over time, indicating that values at time point (t) can be influenced by preceding values at times like $t - 1$ or $t - 2$.

e) **Seasonality**: Recurrent fluctuations at fixed intervals (e.g., daily, monthly, yearly), influenced by factors like time of year, month, or day which are predictable and repetitious. Examples are retail sales increasing during popular holidays.

f) **Trends**: Long-term movement in the data indicates direction and movement over time. Examples are rising global temperatures and housing prices post pandemic.

g) **Cyclic patterns**: Recurrent phenomena without fixed periods, attributed to complex circumstances that are unpredictable and challenging to identify. Examples are forest growth and fire cycles.

h) **Irregularity or noise (irregular component):**
Random variations without a recurring pattern,
attributed to unforeseen events or anomalies.
Examples are rapid stock market fluctuations before
and after a political event.

i) **Frequency:** Data is sampled at regular time
intervals (e.g., hourly, daily, monthly).

j) **Duration:** Length of time between observations.

1.2 Time Series Forecasting Methods

Various techniques and algorithms are available to perform time series
forecasting based on the data characteristics learned in the above section.
They can be "broadly" classified into two categories – univariate and
multivariate.

1.2.1 Univariate

Univariate time series analysis focuses on the study of a single time series
to understand its underlying patterns and make forecasts. Let's understand
some popular techniques:

a) **Moving Average (MA):** The Moving Average model
computes the average of a fixed number of previous
observations to predict future values.

b) **Autoregressive (AR):** Autoregressive models are a
class of models that describe a linear relationship
between an observation at a particular time and a
certain number of lagged observations (i.e., past
values) of the same series.

c) **Autoregressive Moving Average (ARMA)**: This model is a combination of AR (Autoregressive) and MA (Moving Average), and this combination is done to improve the approximation.

d) **Autoregressive Integrated Moving Average (ARIMA)**: This model is a combination of three models – AR (Autoregressive), MA (Moving Average), and Integrated (the number of times differencing is done to make data stationary).

e) **Seasonal Autoregressive Integrated Moving Average (SARIMA)**: SARIMA is an extension of ARIMA that can handle seasonal effects present in the data.

f) **Exponential Smoothing**: Exponential smoothing methods forecast future values by weighting past observations exponentially.

g) **SES**: Suitable for data without trend or seasonality.

h) **Holt's Linear Trend Model**: Extends SES to capture linear trends.

i) **Holt-Winters Seasonal Model**: Extends Holt's model to capture seasonality.

j) **Fourier Analysis**: Fourier Analysis decomposes a time series into sinusoidal components. It is useful for identifying cyclical patterns.

k) **Kalman Filter**: The Kalman filter is an algorithm that uses a series of measurements over time, containing statistical noise and other inaccuracies, to estimate unknown variables.

l) **Hidden Markov Models**: Models time series data as sequences of hidden states with observable outcomes, useful for sequential data with unknown state transitions.

1.2.2 Multivariate

Multivariate time series analysis extends the techniques used in univariate time series to multiple interrelated time series. Exogenous variables which are external factors affecting the target variable are included to make models robust. Examples are sales of the book impacted by exogenous variables such as target audience, reviews, and current topics in trend.

a) **Seasonal Autoregressive Integrated Moving Average with Exogenous Regressors (SARIMAX)**: SARIMAX is an extension of ARIMA which can handle seasonal effects and also include external influencing factors into the model.

b) **Vector Autoregression (VAR)**: VAR models generalize the univariate autoregressive model to capture the linear interdependencies among multiple time series.

c) **Vector Autoregressive Moving Average (VARMA)**: VARMA models extend VAR models by including moving average terms.

d) **Vector Autoregression Moving Average with Exogenous Regressors (VARMAX)**: This model is an extended version of VAR and VARMA models by incorporating exogenous variables.

e) **Vector Error Correction Model (VECM)**: VECM is used for nonstationary time series that are cointegrated. It extends the VAR model to include error correction terms, capturing long-term equilibrium relationships.

f) **Generalized Autoregressive Conditional Heteroskedasticity Models (GARCH)**: GARCH models are designed to capture the changing variances over time, especially useful for modeling financial time series data which often exhibit volatility clustering which are periods of oscillation followed by a period of relative calm.

g) **Convolutional Neural Networks (CNNs)**: CNNs can be adapted to capture spatial dependencies in multivariate time series by treating time series data as images or sequences.

h) **Recurrent Neural Network (RNN), Gated Recurrent Unit (GRU), Long Short-Term Memory (LSTM)**: A type of neural network that is well suited for sequence prediction problems. These neural networks can capture long-term dependencies in multivariate time series.

i) **Transformers**: Originally developed for natural language processing, transformers can be adapted for multivariate time series by capturing relationships between different variables and leveraging attention mechanisms.

Note   Those who are completely new to time series and are interested in understanding more about the above techniques can refer to book [1].

1.3 Introduction to Generative AI

"Even with its very limited current capability and its very deep flaws, people are finding ways to use [AI tools] for great productivity gains or other gains and understand the limitations."

—Sam Altman, CEO of OpenAI

"Some people call this artificial intelligence, but the reality is this technology will enhance us. So instead of artificial intelligence, I think we'll augment our intelligence."

—Ginni Rometty, Former CEO of IBM

"The transformation opportunity that AI brings for all of society, for governments, business, communities, and just human beings, can only be achieved if we have strong public and private sector collaboration."

—Sabastian Niles, President and Chief Legal Officer at Salesforce

These recent quotes from industry leaders highlight the excitement and transformative potential of AI, particularly in its evolving forms like generative AI. Generative AI is a subset of artificial intelligence because it utilizes AI techniques, such as machine learning and pattern recognition, to generate new content, like images and text; just as how a painter uses brushes to create art, GenAI uses algorithms to create new content, making it a specialized tool within the broader scope of AI. For example, ChatGPT, a GenAI tool, uses AI algorithms to generate human-like text responses, making it a subset of AI.

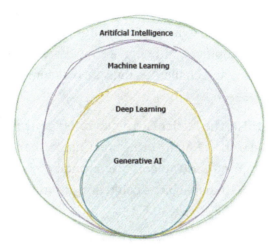

Figure 1-1. *AI and its subsets*

GenAI involves leveraging AI to generate novel content, such as text, images, music, audio, and videos, by employing machine learning algorithms to identify patterns and relationships within human-created content. These learned patterns are then used to create new content, effectively mimicking human creativity.

The emergence of GenAI has significant implications for language teaching and learning, which plays a vital role in today's globalized world. Language proficiency enables individuals to communicate effectively, express ideas clearly, and navigate diverse cultural contexts.

1.4 Evolution from AI to Generative AI

Current generative artificial intelligence is still basic. Artificial intelligence (AI) has seen rapid progress, inching us closer to a future where machines exhibit creative capabilities. A prominent branch of AI, known as generative AI, involves algorithms and models that can produce original content, such as images, music, text, and even entire virtual environments.

Despite the impressive advancements in generative AI, it remains in a relatively early stage – akin to a first draft or initial version compared to its potential future development. Although it showcases remarkable abilities, numerous limitations and challenges must be overcome before generative AI can fully realize its potential.

> **1940–1950**: The **birth of artificial intelligence** (AI) with the works of Alan Turing and Claude Shannon, who proposed early models of computation based on the idea that machines could one day mimic human intelligence.

> **1951–1960**: The **Dartmouth Summer Research Project** on artificial intelligence is considered the **birth of AI as a field of study**. Noam Chomsky released *Syntactic Structures*, a book that lays out a style of grammar called "Phase-Structure Grammar," which translates natural language into a format that computers can understand and use.

> **1961–1970**: Joseph Weizenbaum developed the **first chatbot**, **ELIZA**, which can simulate a conversation with a human by using a simple algorithm to generate text responses to questions.

> **1980–1990**: **Neural networks** are developed, which can learn and remember patterns, providing a model for understanding human memory.

2000–2010: **Generative AI** begins to gain momentum, thanks to advancements in machine learning and deep learning, which enable the creation of neural networks that can process and learn from data like the human brain.

2011–2013: Apple releases **Siri**, an AI and NLP assistant that uses predefined commands to perform actions and answer questions. Deep learning techniques start gaining popularity.

2014: **Generative Adversarial Networks** (GANs) are introduced by Ian Goodfellow and Yoshua Bengio, a class of machine learning frameworks that can generate new data based on a given training set.

2015: **AlexNet** wins ImageNet competition, showcasing deep learning's power. The **attention model** is introduced, which solves the problem of traditional architectures that have to remember an entire input sentence before translation.

2016–2017: **AlphaGo** defeats a human Go champion, demonstrating AI's capabilities. Transformers are introduced, revolutionizing natural language processing.

2018: **GPT-1** is released, marking the beginning of generative AI. The generative pre-training of a language model is republished on OpenAI's website, showing how a generative language model can acquire knowledge and process dependencies unsupervised based on pre-training on a large and diverse set of data.

2019: OpenAI releases the complete version of its **GPT-2** language model, which was trained on a dataset of more than nine million documents.

2020: Transformers become widely adopted in natural language processing. AI-powered chatbots become popular in customer service. **GPT-3 released.**

2021: DALL-E and Midjourney introduce **generative AI for images**.

2022: Stability AI develops **Stable Diffusion**, a deep learning text-to-image model that generates images based on text descriptions. ChatGPT releases **GPT-3.5**, an AI tool that can access data from the Web up to 2021.

2023: **TimeGPT**, the first foundation model for time series forecasting, is released. The **generative AI race begins**, with Microsoft integrating ChatGPT technology into Bing; Google releasing its own generative AI chatbot, **Bard**; and OpenAI releasing **GPT-4**.

2024: MIT launched a working group to explore the **future of work with generative AI**, Runway introduced **Gen-2 for high-quality video production**, Google expanded access to its **Gemini AI** models, and Stanford researchers developed the **SyntheMol** AI model for creating new antibiotics to combat resistant bacteria. Powerful foundation models for time series covered in this book start seeing wider adoption.

1.5 Generative AI with Time Series

Over the last decade, machine learning techniques have gained popularity and shown significant promise. Traditionally, statistical methods have dominated time series analysis and forecasting, such as ARIMA, ETS, MSTL, Theta, and CES, which have been widely utilized across diverse domains for their reliability.

Over the past decade, machine learning models such as XGBoost and LightGBM have gained traction, showing strong performance in competitions and real-world applications. However, the emergence of deep learning has marked a significant paradigm shift in time series analysis. Deep learning methods have garnered popularity in academia and have been increasingly adopted for large-scale industrial forecasting tasks.

Ongoing research in generative artificial intelligence is focused on application to time series data and investigating the potential benefits of foundational models. The foundation models are independently trained on vast time series datasets as a large transformer model. The models are developed to minimize the forecasting error. The model thus developed uses the past data window to forecast the future.

The underlying idea is that attention-based mechanisms effectively capture the diversity of past events, enabling accurate extrapolation of potential future distributions. These developments may herald a new phase in the field, deepening our understanding of temporal data and enhancing the efficiency of forecasting and application in various domains.

Transformers have demonstrated exceptional capability in modeling long-range dependencies and interactions within sequential data, making them highly attractive for time series modeling. Numerous transformer variants have been developed to tackle specific challenges in this domain and have proven successful in applications such as forecasting, anomaly detection, and classification. Notably, addressing seasonality and periodicity remains a crucial aspect of time series analysis.

TimeGPT is the first pre-trained foundation model for time series forecasting that can produce accurate predictions across various domains and applications without additional training. The architecture consists of an encoder-decoder structure with multiple layers, each with residual connections and layer normalization. Finally, a linear layer maps the decoder's output to the forecasting window dimension. The general intuition is that attention-based mechanisms can capture the diversity of past events and correctly extrapolate potential future distributions. This innovation marks a significant breakthrough that paves the way for a new forecasting paradigm. The new techniques discussed above are more accessible, accurate, less time-consuming, and substantially reduce computational complexity.

Ongoing advancements aim to enhance generated content's realism, fidelity, and diversity across various formats, including images, text, audio, and video. This involves developing more advanced generative models, employing innovative training techniques, and establishing superior evaluation metrics to assess output quality.

Few-shot and zero-shot learning advances will enable generative models to tackle new tasks or domains with minimal or no training data, reducing reliance on large annotated datasets and enhancing adaptability.

Ensuring the robustness and security of generative models against adversarial attacks is crucial for their practical deployment. Future research will focus on creating defenses against adversarial manipulation and preventing the malicious use of generative AI. Additionally, developing algorithms that can continuously learn and adapt over time, integrating new data and knowledge while retaining previously learned information, will be essential for sustained use in dynamic real-world environments.

As AI becomes increasingly pervasive, addressing ethical and societal issues such as privacy, bias, fairness, and responsible use of synthetic media will be imperative. This requires collaboration across disciplines and the establishment of ethical guidelines, regulatory frameworks, and accountability measures.

1.6 Introduction to Large Language Models

A large language model (LLM) is a type of model developed by training on massive amounts of data. This enables it to understand and generate responses indistinguishable from human responses. These are especially helpful for tasks like translation, summarization, writing creative content, time series forecasting, and image and video generation.

LLMs have seen significant use in domains such as natural language processing and computer vision. Beyond text, images, and graphics, LLMs present substantial potential for analyzing time series data, benefiting fields such as climate science, IoT, healthcare, traffic management, audio processing, and finance. This survey paper provides an in-depth exploration and a detailed taxonomy of the various methodologies employed to harness the power of LLMs for time series analysis. We address the inherent challenge of bridging the gap between LLMs' original text-based training and the numerical nature of time series data, and we explore strategies for transferring and distilling knowledge from LLMs to numerical time series analysis.

Figure 1-2. *Large language models have recently been applied for various time series tasks in diverse application domains from the "Large Language Models for Time Series: A Survey" paper [3]*

In the following chapters, we'll explore high-level theoretical concepts that will provide enough insights to follow them with simple practical implementation.

1.7 Summary

In this chapter, we started with an introduction to time series analysis by understanding its characteristics and various forecasting methods, followed by a deep dive into evolution AI, followed by a gentle introduction to generative AI and large language models.

Bull Whip Effect

This is a phenomena noticed in the supply chain. The orders placed with the manufacturer tend to have a larger variability than sales to end customers. This results in inaccurate demand projections to the manufacturer or upstream supplier.

Cart Loading

This is a phenomena noticed in a retailer's supply chain. During sales like Thanksgiving, customers tend to buy additional quantities of items with a higher shelf life than their regular shopping habits. This is due to discounts and offers on items. This results in challenges with retailer's estimates of quantities to stock. This occurs due to customers' changes in shopping habits for the next couple of months or visits.

1.8 References

[1].   Hands-on Time Series Analysis with Python: From Basics to Bleeding Edge Techniques by B V Vishwas (Author), Ashish Patel. https://doi.org/10.1007/978-1-4842-5992-4

[2].   TimeGPT, Azul Garza, Cristian Challu, Max Mergenthaler-Canseco. Nixtla San Francisco, CA, USA. https://doi.org/10.48550/arXiv.2310.03589

[3].   Transformers in Time Series: A Survey. https://doi.org/10.48550/arXiv.2202.07125

CHAPTER 2

Neural Networks for Time Series

Chapter Goal: Learn how to leverage different types of neural network architectures to solve time series problems.

In the previous chapter, we understood the evolution of artificial intelligence and covered the basics of time series and the introduction to generative AI and large language models.

In this chapter, let us understand techniques related to time series analysis using neural networks. We will focus on simple perceptron, multilayer perceptron, convolutional neural network, recurrent neural network, long short-term memory, and autoregressive and neural basis expansion analysis for interpretable time series.

2 Introduction to Perceptron

In the following sections, let us cover some techniques related to time series analysis using neural networks. We will be building up on some basics here before working with foundation models. Foundation models are also a type of neural network models. The first model that comes to our mind is a perceptron. Let us understand what perceptron and multilayer perceptron are and implement a use case.

© Banglore Vijay Kumar Vishwas and Sri Ram Macharla 2025
B. V. Vishwas and Sri Ram Macharla, *Time Series Forecasting Using Generative AI*,
https://doi.org/10.1007/979-8-8688-1276-7_2

A perceptron may be understood as the simplest form of neural network. It is a type of neural network with a single neuron. The perceptron algorithm is among one of the earliest algorithms used for supervised learning.

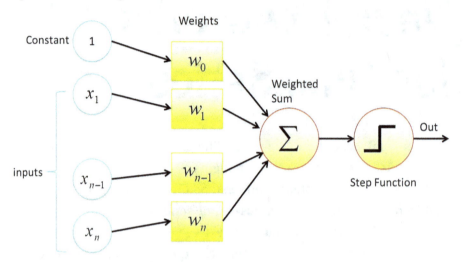

Figure 2-1. *Perceptron example*

The intuition behind the working of a perceptron is that it accepts several binary inputs; each input is multiplied by a weight. Finally, add all the weighted inputs. This value is passed through a step function. This results in a single binary output.

The step function results in one if the input is greater than or equal to zero, and zero for all other inputs. Hence, the step function is useful for binary classification. This function is used in threshold-based models and not in the basic perceptron.

2.1 Technical Overview of a Perceptron

As a first step, the perceptron receives inputs. The inputs could be independent variables/features. These inputs are combined (multiplied) with a set of weights. The perceptron's formula can be expressed as

$$\text{output} = f\left(w_1{}^* \, x_1 + w_2{}^* \, x_2 + \ldots + w_n{}^* \, x_n + b\right) \qquad \text{(Eq. 2.1)}$$

where

$w_1, w_2, \ldots, w_n$ are the weights.

$x_1, x_2, \ldots, x_n$ are the input signals.

b is the bias, which allows the activation function to be shifted to the left or right, to better fit the data.

f is the activation function, typically a step function that outputs either zero or one.

The perceptron's decision-making process is binary. If the sum of the weighted inputs plus the bias is greater than zero, the perceptron outputs a one; otherwise, it outputs a zero. This binary step function is what allows the perceptron to classify input data.

From the above equation, we can understand that weights are a set of values associated with the connections between neurons. They determine the strength of these connections. They control the influence that one neuron's output has on another neuron's input. Weights may be understood as the coefficients of the input variables that adjust the impact of incoming data. They can increase or decrease the importance of specific information.

During the training phase of a neural network, the weights are adjusted in iterations. This helps in reducing the difference between the model's prediction and the actual outcomes to a minimum. This process helps in fine-tuning the network's ability to make accurate predictions.

Weights are the neural network's mechanism to learn from data. Weights capture the relationships between input features and the target output feature. This allows the network to generalize and make predictions on new, unseen data.

In Equation 2.1, we can see a value b added at the end. This bias value is a constant associated with each neuron. Unlike weights, biases are not combined with specific inputs but are added to the neuron's output. Bias serves as a form of offset or threshold, helping neurons to activate even when the weighted sum of their inputs is not sufficient on its own. They introduce a level of adaptability that ensures the network can learn and make predictions effectively.

The result of the weighted sum plus bias is passed through an activation function. This function determines whether the neuron should activate or remain dormant based on the calculated value.

While training the neural network, the values of weights and bias are adjusted through an optimization process. The most frequently used technique is named gradient descent, and it is used along with a learning algorithm called backpropagation. Using this gradient descent optimization method, the gradient of the error is computed.

This computation is performed with respect to the values of weights and bias. The gradient of the error is nothing but difference between the predicted value and the actual value. This gradient points toward the steepest decrease in error. The neural network updates the values of weights and bias in small steps. The intention is minimizing the error. This entire process is repeated until the neural network reaches a state where the prediction error is minimal.

Now a question may arise – what are the starting values of weights?

Before the start of the training, weights in an ANN (Artificial Neural Network) must be initialized to some values. Proper weight initialization plays a key role in the convergence and performance of the network. The most common initialization method is random initialization. As the name says, the weights are assigned small random values.

2.2 What Is Multilayer Perceptron?

In the previous section, we learned how a perceptron works. In this section, let us understand the multilayer perceptron (MLP).

An MLP is a neural network that has at least three layers: an input layer, a hidden layer, and an output layer. Each layer performs operations on the outputs of its preceding layer.

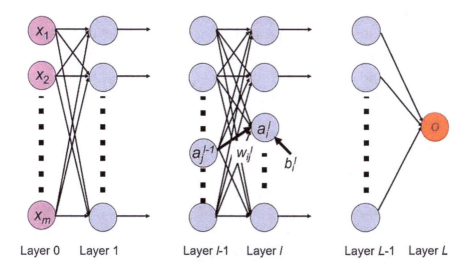

Layer 0   Layer 1   Layer l-1   Layer l Layer L-1   Layer L

Figure 2-2. *Multilayer perceptron*

In Figure 2-2, we use the following notations:

a_i^l is the activation (output) of neuron i in layer l.

w_{ij}^l is the weight of the connection from neuron j in layer l-1 to neuron i in layer l.

b_i^l is the bias term of neuron i in layer l.

The hidden layers are nothing but intermediate layers between the input and the output layers.

Now that we have understood simple neural networks, let us explore CNN-based architecture and how it can be leveraged for time series forecasting.

For time series analysis in this book, one of the libraries that will be used is the NeuralForecast library. It is a library for time series forecasting developed in Python.

This library has built-in datasets, statistical tests, benchmarks, utilities for evaluation, and data loading. There are many reasons to use this library – fast and accurate implementations of state-of-the-art models, support for exogenous variables and static covariates, probabilistic forecasting, and more.

You can find more details at `https://github.com/Nixtla/neuralforecast.`

Note   Extra attention should be given while setting up the Python package for the code to work in each notebook.

2.2.1 Multilayer Perceptron in Action

Having established a high-level theoretical foundation of MLP, we shall now translate abstract concepts into practical code implementation.

Help an airline company to estimate the number of air passengers per month. Let's leverage a dataset with a monthly passenger count for 12 years. This dataset is used to train the model and then forecast the passenger traffic for the subsequent year.

Note   The same dataset is used throughout the book for most of the univariant examples.

Import libraries:

```
import pandas as pd
import numpy as np
from sklearn.metrics import mean_squared_error, mean_absolute_
percentage_error, r2_score
from IPython.display import display, Markdown
import matplotlib.pyplot as plt
from neuralforecast import NeuralForecast
from neuralforecast.losses.pytorch import MQLoss,
DistributionLoss
from neuralforecast.models import MLP
```

Define the calculate_error_metrics function which helps in assessing the performance of the trained model.

Note   This function is referred to throughout the book.

```
def calculate_error_metrics(actual, predicted, num_
predictors=1):
    # convert inputs are numpy arrays
    actual = np.array(actual)
    predicted = np.array(predicted)
    # Number of observations
    n = len(actual)
    # Calculate MSE
    mse = mean_squared_error(actual, predicted)
    # Calculate RMSE
    rmse = np.sqrt(mse)
    # Calculate MAPE
    mape = mean_absolute_percentage_error(actual, predicted)
    # Calculate R-squared
```

```
r2 = r2_score(actual, predicted)
# Calculate Adjusted R-squared
adjusted_r2 = 1 - ((1 - r2) * (n - 1) / (n - num_
predictors - 1))
print(f'MSE : {mse}')
print(f'RMSE : {rmse}')
print(f'MAPE : {mape}')
print(f'r2 : {r2}')
print(f'adjusted_r2 : {adjusted_r2}')
```

Let's load the dataset either from an offline copy or from the
neuralforecast.utils dataset, which contains 12 years of monthly air
passenger count. Separate the last 1 year of data for the test and use the
remaining 11 years of data to train the model.

```
from neuralforecast.utils import AirPassengersDF
Y_df = AirPassengersDF
Y_df = Y_df.reset_index(drop=True)
Y_df.head()
```

	unique_id	ds	y
0	1.0	1949-01-31	112.0
1	1.0	1949-02-28	118.0
2	1.0	1949-03-31	132.0
3	1.0	1949-04-30	129.0
4	1.0	1949-05-31	121.0

```
train_data = Y_df.head(132)
test_data = Y_df.tail(12)
```

Let's initialize and train the multilayer perceptron by understanding key parameters.

h is the forecast horizon.

input_size is considered the autoregressive inputs (lags), y=[1,2,3,4] input_size=2 -> lags=[1,2].

Loss is PyTorch module, instantiated train loss class from the losses collection.

scaler_type is the step size between each window of temporal data.

learning_rate is the learning rate between (0, 1).

max_steps is the maximum number of training steps.

val_check_steps is the number of training steps between every validation loss check.

early_stop_patience_steps is the number of validation iterations before early stopping.

```
horizon = 12
model = MLP(h=horizon, input_size=12,
            loss=DistributionLoss(distribution='Normal',
            level=[80, 90]),
            scaler_type='robust',
            learning_rate=1e-3,
            max_steps=200,
            val_check_steps=10,
            early_stop_patience_steps=2)

fcst = NeuralForecast(models=[model],freq='M')
fcst.fit(df=train_data, val_size=12)
```

Predict the next defined horizon:

```
Y_hat_df = fcst.predict()
Y_hat_df.head()
```

The MLP column contains the predicted values as depicted below:

unique_id	ds	MLP	MLP-median	MLP-lo-90	MLP-lo-80	MLP-hi-80	MLP-hi-90
1.0	1960-01-31	408.924683	408.840027	386.458618	391.899323	425.048157	429.648743
1.0	1960-02-29	392.037598	391.304260	361.825439	368.233765	416.101013	424.014893
1.0	1960-03-31	484.258087	484.340729	437.918213	448.213684	519.549988	531.080200
1.0	1960-04-30	468.928833	469.907593	420.667877	431.899353	505.866058	515.705750
1.0	1960-05-31	464.647125	466.234467	413.373505	424.859863	503.730225	512.233765

Measure the model's accuracy:

```
calculate_error_metrics(test_data[['y']],Y_hat_df['MLP'])
```

```
MSE : 547.1603393554688
RMSE : 23.39145851135254
MAPE : 0.03406880423426628
r2 : 0.9012251869626534
adjusted_r2 : 0.8913477056589187
```

Visualize the predictions:

```
train_data.set_index('ds',inplace =True)
Y_hat_df.set_index('ds',inplace =True)
test_data.set_index('ds',inplace =True)
plt.figure(figsize=(20, 3))
y_past = train_data["y"]
y_pred = Y_hat_df['MLP']
y_test = test_data["y"]
plt.plot(y_past, label="Past time series values")
plt.plot(y_pred, label="Forecast")
plt.plot(y_test, label="Actual time series values")
plt.title('AirPassengers Forecast', fontsize=10)
```

```
plt.ylabel('Monthly Passengers', fontsize=10)
plt.xlabel('Timestamp [t]', fontsize=10)
plt.legend();
```

Figure 2-3. *Actual vs. predicted plot*

Figure 2-3 helps us to appreciate that the air passenger count predicted by our model is very close to reality.

2.3 CNN-Based Architecture for Time Series

Convolutional neural networks are a type of neural nets best suited for computer vision and speech processing. This type of network has a minimum of three layers – convolutional layer, pooling layer, and fully connected layer. In the convolutional layer, the features are extracted by applying convolutional filters while retaining the spatial relationship between pixels. The operations performed in this layer result in dimensionality reduction without impacting essential features. These feature maps are passed on to the pooling layer. In the pooling layer, downsampling is done on the feature maps, which results in reducing their spatial dimensions while retaining essential features. This helps in reducing overfitting and makes the model immune to small changes. In a fully connected layer, the final classification task is performed based on extracted features from the previous layers. CNN architectures are primarily used in image processing applications in the medical domain [5], industrial automation, quality control, autonomous driving, and time series.

In this section, we will be covering different CNN-based forecasting techniques like WaveNet, TCN (temporal convolutional network), and BiTCN (bidirectional temporal convolutional network).

2.3.1 WaveNet for Time Series Forecasting

Let us explore WaveNet architecture, a deep neural network that can be used for time series forecasting. WaveNet was primarily developed for music and audio generation. WaveNet DNN may be classified as a generative model which is based on a dilated causal convolutional neural network. Let us explore modifying the WaveNet model for time series forecasting. In order to learn long-term dependencies with the time series data, it uses stacked layers of dilated convolutions.

Dilated convolutions allow WaveNet to efficiently learn long-range relationships in the data without sacrificing computational efficiency. WaveNet's DNN structure is designed in a manner that the model only uses past values to predict future values (causality), all this while keeping intact the temporal dependencies of the data. Temporal dependency involves the impact of previous behavior on current behavior. Temporal dependencies are relationships between past and future events in a time series data. They can be useful for predicting outcomes and understanding patterns.

2.3.2 Technical Overview of WaveNet

The WaveNet architecture uses a combination of complex mathematical operations to generate and model sequences. Understanding the math behind WaveNet architecture requires deeper understanding of its three constituent components: (a) dilated convolutions, (b) causal convolutions, and (c) residual connections. Let's understand the math behind individual components of WaveNet:

a) **Dilated Convolutions**

Dilated convolutions help the model to have a larger receptive field. The interesting part is that this is achieved without increasing the number of parameters or the computational complexity significantly.

Given a 1D convolution operation with a kernel K and dilation rate, the dilated convolution operation can be expressed as

$$y(t) = \sum_{i=0}^{K-1} w(i) \cdot x(t - d \cdot i)$$

where:

- $y(t)$ is the output at time t.

- $w(i)$ is the weight at position i in the kernel.

- $x(t - d \cdot i)$ is the input at position $t - d \cdot i$.

- d is the dilation rate.

- K is the size of the kernel.

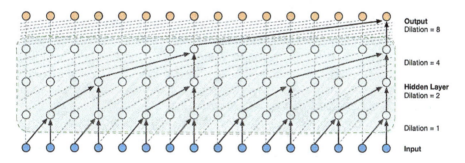

Figure 2-4. *Dilated convolution with a dilation rate of 2*

The dilation rate d effectively expands the kernel's receptive field. For example, a dilation rate of 2 means that the kernel will have gaps of 1 between each weight. This helps the model to cover a larger input span with fewer parameters.

29

b) **Causal Convolutions**

Causal convolutions ensure that the model does not use future values to predict past values. This is necessary for time series and sequence prediction tasks.

For a causal convolution, the output at time t only depends on the current and past inputs. This is achieved by padding the input sequence with zeros at the beginning. Mathematically, it is expressed as

$$y(t) = \sum_{i=0}^{K-1} w(i) \cdot x(t-i)$$

where:

- $x(t-i)$ is zero for $t-i < 0$.

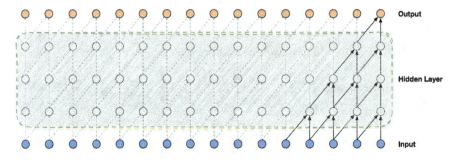

Figure 2-5. *Depiction of casual convolution*

c) **Residual Connections**

Residual connections help in training the neural network by mitigating the vanishing gradient problem. They also help the gradients to flow through the network more effectively.

$$y(t) = F(x(t)) + x(t)$$

where:

- $F(x(t))$ is the output of the convolutional block.
- $x(t)$ is the input to the block.

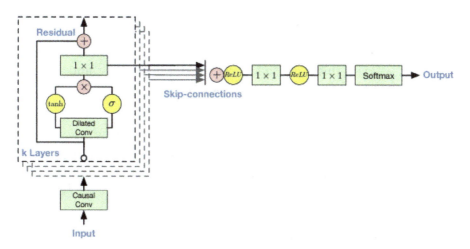

Figure 2-6. *Depiction of residual block and complete architecture [4]*

Learning of residual mapping by the model is taken care of by residual connection. This is achieved by adding the input to the output of the convolution block.

The WaveNet model achieves a large receptive field by deploying multiple dilated causal convolutions. The residual connections help in building deeper networks. They also improve the gradient flow. Gated activations help capture complex temporal patterns by introducing nonlinearity and enhancing modeling capabilities. For audio generation, the final output layer typically uses a softmax function. This helps to generate probabilities for the next value in the sequence.

In order to leverage the WaveNet DNN model for time series forecasting, the input needs to be a time series data instead of waveform. The input data could be original values, derived feature values, or

normalized values. This can be used to predict the next time step or
series of future values. The preferred loss function is MSE (mean squared
error) or MAE (mean absolute error). To determine the optimal WaveNet
architecture for our particular time series dataset, it is necessary to
perform hyper-parameter tuning, specifically adjusting the layers,
filters, and dilation rate. In real-world applications, we were able to
improve the results of predictions by combining the WaveNet model with
autoregressive models.

Consider using this model when computation, memory, and power
consumption are not a challenge. You should also be having access
to a large amount of data. This model shines when there are temporal
dependencies in the data. This model handles most of the types of time
series data and can be adapted to various time series forecasting use
cases. One of the downsides of this model is that it is challenging to tune
hyperparameters.

2.3.3 WaveNet in Action

AutoGluon is an open source library for automating machine learning
tasks. This library supports training and deployment of ML and deep
learning models. It provides support for time series forecasting models
which are used throughout the book for implementing few techniques.
Please refer to [1] for more details.

Now that we have understood the theoretical aspects of WaveNet, let
us see a practical implementation.

Import libraries:

```
import autogluon
from neuralforecast.utils import AirPassengersDF
import numpy as np
from sklearn.metrics import mean_squared_error, mean_absolute_
percentage_error, r2_score
from autogluon.timeseries import TimeSeriesPredictor,
```

```
TimeSeriesDataFrame
from autogluon.timeseries.models import WaveNetModel
import pandas as pd
```

Let's load the dataset either from an offline copy or from the **neuralforecast.utils** dataset, which contains 12 years of monthly air passenger count. Separate the last 1 year of data for the test and use the remaining 11 years of data to train the model.

```
Y_df = AirPassengersDF
Y_df = Y_df.reset_index(drop=True)
Y_df.head()
```

	unique_id	ds	y
0	airline_1	1949-01-31	112
1	airline_1	1949-02-28	118
2	airline_1	1949-03-31	132
3	airline_1	1949-04-30	129
4	airline_1	1949-05-31	121

Convert the unique_id column to categorical so that we can convert to the format which AutoGluon understands:

```
Y_df['ds'] = pd.to_datetime(Y_df['ds'])
Y_df['unique_id'] = 'airline_1'
```

AutoGluon expects time series data in long format. Each row of the data frame contains a single observation (time step) of a single time series represented by

- Unique ID of the time series item_id as int or str

- Timestamp of the observation timestamp as a pandas. Timestamp

- Compatible format
- Numeric value of the time series target

```
data = TimeSeriesDataFrame.from_data_frame(
    Y_df,
    id_column="unique_id",
    timestamp_column="ds"
)
data.tail()
```

		y
item_id	timestamp	
airline_1	1960-08-31	606
	1960-09-30	508
	1960-10-31	461
	1960-11-30	390
	1960-12-31	432

Split data into train and test:

```
train_data = data.head(132)
test_data = data.tail(12)
```

Create a TimeSeriesPredictor object to forecast future values and explicitly define a WaveNet model to be used:

```
predictor = TimeSeriesPredictor(target='y',
                                prediction_length=12,
                                eval_metric="MASE",).fit
(train_data,presets='best_quality', hyperparameters={'WaveNetMo
del': {}},time_limit=600)
```

Predict the next defined horizon:

```
predictions = predictor.predict(train_data)
predictions.head()
```

The mean is the predicted column of interest.

item_id	timestamp	mean	0.1	0.2	0.3	0.4	0.5	0.6	0.7	0.8	0.9
airline_1	1960-01-31	487.297943	410.974152	426.630341	448.940353	465.770691	487.297943	508.825134	543.660089	567.535706	610.590149
	1960-02-29	493.169006	422.324844	440.720856	469.684723	483.775299	493.169006	520.567261	547.965515	568.318518	610.590149
	1960-03-31	485.340912	422.716248	438.372437	452.854385	465.770691	485.340912	506.476715	529.569586	555.793640	595.325427
	1960-04-30	489.254974	406.668716	426.630341	454.028595	470.467529	489.254974	504.911102	542.485870	576.146594	602.762085
	1960-05-31	479.469864	410.974152	436.806824	452.854385	465.770691	479.469864	494.734619	515.087616	561.273254	600.022290

Measure the model's accuracy:

```
calculate_error_metrics(test_data['y'],predictions['mean']
['airline_1'].tail(48))
```

```
MSE : 5787.434110611172
RMSE : 76.07518722034914
MAPE : 0.1376407025909576
r2 : -0.04476273242430562
adjusted_r2 : -0.1492390056667361
```

Visualize the predictions:

```
import matplotlib.pyplot as plt
plt.figure(figsize=(20, 3))
item_id = "airline_1"
y_past = train_data.loc[item_id]["y"]
y_pred = predictions.loc[item_id]
y_test = test_data.loc[item_id]["y"]
plt.plot(y_past, label="Past time series values")
plt.plot(y_pred["mean"], label="Mean forecast")
plt.plot(y_test, label="Actual time series values")
plt.title('AirPassengers Forecast', fontsize=10)
```

```
plt.ylabel('Monthly Passengers', fontsize=10)
plt.xlabel('Timestamp [t]', fontsize=10)
plt.fill_between(
    y_pred.index, y_pred["0.1"], y_pred["0.9"], color="red",
alpha=0.1, label=f"10%-90% confidence interval"
)
plt.legend();
```

Figure 2-7. *Actual vs. predicted plot*

Figure 2-7 helps us to appreciate that the air passenger count predicted by our model is not close to reality. This model is best used to capture complex relationships within the signals like in the case of electroencephalogram time series data.

2.4.1 Temporal Convolutional Networks

The temporal convolutional network (TCN) is a type of convolutional neural network architecture best suited for use cases involving sequential data. The models built on these architectures work by exploiting the capabilities of convolution operations mapped to temporal dimension. This helps in learning patterns in the sequential data and also captures long-range dependencies.

Temporal dimension is the dimension in a dataset that represents progression in time. In our area of interest, i.e., the time series data, it can be identified by looking at columns in which observations are ordered chronologically. For example, in the dataset of air travel, the year represents a temporal dimension. In time series forecasting, the temporal dimension helps to understand the effects of past observations on future values.

In machine learning models like the CNN, 1D convolutions can be used to capture local patterns over time. This is achieved by learning relationships between elements at progressive time intervals. In the WaveNet architecture discussed in the previous section, dilated convolutions were used to extend the receptive field. This helps to capture longer-term dependencies without increasing the computational load.

Before the advent of TCNs, an approach where CNNs were combined with RNNs was used. CNNs helped to capture spatial relationships, while RNNs helped to capture temporal relationships. However, with the addition of GPU and TPU processors, TCNs can capture spatiotemporal relations simultaneously with high degree of parallelism. Convolution operations can be performed in parallel, making TCNs more efficient than recurrent networks. Remember that RNNs are inherently sequential networks.

2.4.2.1 Technical Overview of TCN

In Figure 2-8, we can see a TCN with multiple layers, each corresponding to exponentially increasing dilation factors d = 1, 2, 4. The input layer is represented by blue circles, the hidden layer is represented by red circles, and the output layer is represented by yellow circles.

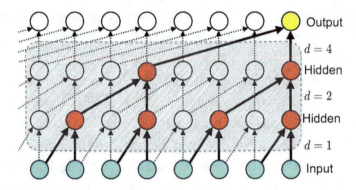

Figure 2-8. *Architecture of TCN [6]*

The advantage of using TCN is that this model can capture longer dependencies compared to LSTM or RNN. However, it may be noted that newer models like transformers have much better range. One of the downsides of using this model is that TCNs could be computationally expensive, as seen in long sequences. Let's break it down further and understand.

Input

The input could be thought of as a matrix of dimensions L X F where L and F are the length and features of the input time series dataset.

Convolution Layer

The convolution layer applies a filter W with dimensions $K \, x \, F$ to the time series data. Here, K is the kernel size, F is the number of features, and the length of stride is denoted by S. Convolution between filters (each filter) and the input sequence results in an output sequence of length T which is the length of the series.

Padding

This ensures that the output sequence has the same length as the input sequence. Zero padding adds zeros to the beginning and end of the input sequence.

Dilated Convolutions

These help the filter to capture long-range temporal dependencies. This is done by increasing the spacing between filter weights. The spacing is controlled by dilation factor D.

Activation Function

The output of each convolution layer is passed through an activation function, such as ReLU or sigmoid, to introduce nonlinearity.

Residual Connections

These allow information to flow directly from previous layers to later layers. This helps to prevent vanishing gradients and improve training stability.

Pooling

Pooling layers reduce the dimensionality of the feature maps by selecting the maximum or average value within a sliding window. This helps to extract salient features and reduce computational costs.

Output Layer

The TCN outputs a prediction for each time step. For our task of forecasting, we could use a linear function for the output layer.

$$\hat{y}(T + 1) = W \cdot H(T) + b$$

where

- W is the weight matrix of the final linear layer.

- b is the bias vector.

- H(T) is the feature vector from the last convolution block.

- $\hat{y}$ (T+1) is the predicted value for the next time step.

Loss Function

This measures the difference between the predicted output and the actual output. For our task, we could use the mean squared error (MSE).

Optimization

The model parameters (weights and biases) are computed using gradient descent algorithms like Adam or SGD (Stochastic Gradient Descent). This helps minimize the loss function.

Another important concept to know about the TCN is the input receptive field (IRF). IRF is the maximum time span for which a single output neuron can receive information from the input sequence.

The input receptive field of a TCN is dependent on three parameters: (a) convolution kernel size, (b) number of hidden layers, and (c) dilation factor.

The predictions are obtained by transforming the hidden states into contexts $\mathbf{c}_{[t+1:t+H]}$, which are decoded and adapted into $\hat{y}_{[t+1:t+H],\,[q]}$ through MLPs.

where $\mathbf{h}_t$ is the hidden state for time t, $\mathbf{y}_t$ is the input at time t, $\mathbf{h}_{t-1}$ is the hidden state of the previous layer at $t-1$, $\mathbf{x}^{(s)}$ are static exogenous inputs, $\mathbf{x}_t^{(h)}$ are historic exogenous, and $\mathrm{x}_{[:t+H]}^{(f)}$ are future exogenous available at the time of the prediction.

2.4.2.2 TCN in Action

Having established a high-level theoretical foundation of TCN, we shall now translate abstract concepts into practical code implementation.

Import libraries:

```
import pandas as pd
import numpy as np
from sklearn.metrics import mean_squared_error, mean_absolute_
percentage_error, r2_score
from IPython.display import display, Markdown
import matplotlib.pyplot as plt
from neuralforecast import NeuralForecast
from neuralforecast.losses.pytorch import GMM, MQLoss,
DistributionLoss
```

```
from neuralforecast.auto import TCN
from neuralforecast.tsdataset import TimeSeriesDataset
from ray import tune
```

Load the AirPassenger dataset and split data into train and test:

```
from neuralforecast.utils import AirPassengersDF as Y_df

Y_train_df = Y_df[Y_df.ds<='1959-12-31']
Y_test_df = Y_df[Y_df.ds>'1959-12-31']
dataset, *_ = TimeSeriesDataset.from_df(Y_train_df)
```

Let's initialize and train the model (TCN) by understanding its key parameters.

h is the forecast horizon.

input_size is the maximum sequence length for truncated train backpropagation. Default –1 uses all history.

Loss is the instantiated train loss class.

learning_rate is the learning rate between (0, 1).

kernel_size is the size of the convolving kernel.

dilations controls the temporal spacing between the kernel points, also known as the à trous algorithm.

encoder_hidden_size is the unit for the TCN's hidden state size.

context_size is the size of the context vector for each timestamp on the forecasting window.

decoder_hidden_size is the size of the hidden layer for the MLP decoder.

decoder_layers is the number of layers for the MLP decoder.

max_steps is the maximum number of training steps.

scaler_type is the type of scaler for temporal input normalization; see temporal.

hist_exog_list is the historic exogenous columns.

```
horizon = 12
fcst = NeuralForecast(
    models=[TCN(h=horizon,
                input_size=-1,
                loss=GMM(n_components=7, return_params=True,
                level=[80,90]),
                learning_rate=5e-4,
                kernel_size=2,
                dilations=[1,2,4,8,16],
                encoder_hidden_size=128,
                context_size=10,
                decoder_hidden_size=128,
                decoder_layers=2,
                max_steps=500,
                scaler_type='robust',
                #futr_exog_list=['y_[lag12]'],
                hist_exog_list=None,
                #stat_exog_list=['airline1'],
                )
    ],
    freq='M'
)

fcst.fit(df =Y_train_df)
```

Predict for the next defined horizon:

```
y_hat = fcst.predict()
y_hat.set_index('ds',inplace =True)
y_hat.head()
```

TCN is the predicted column of interest.

	TCN	TCN-median	TCN-lo-90	TCN-lo-80	TCN-hi-80	TCN-hi-90	TCN-mu-1	TCN-std-1	TCN-mu-2	TCN-std-2	TCN-mu-3	TCN-std-3	TCN-mu-4
ds													
1960-01-31	399.866486	400.442200	370.854034	378.432098	421.345367	427.704620	404.990723	15.600019	391.500183	15.600014	409.794250	15.600022	397.643433
1960-02-29	408.066925	408.250916	372.417664	380.781403	436.946503	443.140747	412.740601	15.600007	386.670105	15.600008	395.573120	15.600008	421.659851
1960-03-31	431.713684	433.548218	395.079865	403.510925	456.927521	464.503021	443.772888	15.600001	435.004578	15.600010	450.948792	15.600006	433.966583
1960-04-30	419.090576	421.557922	379.846100	385.572418	447.335663	454.024139	429.013062	15.600054	431.354309	15.600049	427.829102	15.600060	435.109619

Measure the model's accuracy:

```
calculate_error_metrics(Y_test_df[['y']],y_hat[['TCN']])
```

```
MSE : 433.3346862792969
RMSE : 20.816692352294922
MAPE : 0.03554682061076164
r2 : 0.9217732930276429
adjusted_r2 : 0.9139506223304071
```

Visualize the predictions:

```
Y_train_df.set_index('ds',inplace =True)
Y_test_df.set_index('ds',inplace =True)
plt.figure(figsize=(20, 3))
y_past = Y_train_df["y"]
y_pred = y_hat[['TCN']]
y_test = Y_test_df["y"]
plt.plot(y_past, label="Past time series values")
plt.plot(y_pred, label="Forecast")
plt.plot(y_test, label="Actual time series values")
plt.title('AirPassengers Forecast', fontsize=10)
plt.ylabel('Monthly Passengers', fontsize=10)
plt.xlabel('Timestamp [t]', fontsize=10)
plt.legend();
```

Figure 2-9. *Actual vs. predicted plot*

Figure 2-9 helps us to appreciate that the air passenger count predicted by our model is close to reality.

2.4.3 Bidirectional Temporal Convolutional Network

In the previous section, we understood the working of TCN. In this section, let us understand a CNN architecture that leverages the predictive power of combining two TCN networks.

While approaches that follow transformer architecture (covered in upcoming chapters) deliver cutting-edge performance, it comes at the cost of high compute and memory. This is due to the fact that transformer-based approaches learn a large number of parameters. The BiTCN handles this challenge by combining two TCNs. Bidirectional temporal convolutional network (BiTCN) architecture is developed by using two temporal convolutional networks for forecasting. The first network, called the forward network, encodes future covariates of the time series. The second network, called the backward network, encodes past observations and covariates. This technique helps in preserving temporal information of sequence data. The parameters of the output distribution are jointly estimated using these two networks. It is computationally more efficient than RNN methods like LSTM. Compared to newer architectures like transformer-based methods, it requires parameters of lower order magnitude (lower space complexity). BiTCN falls under the category of univariate models.

The benefits of a lesser number of parameters directly translate to lower memory and computing costs, besides lower cost of deployment. Choose this model when you are looking for a model with a lesser number of hyperparameters to tune and a smaller number of trainable parameters.

2.4.3.1 What Are Future Covariates?

The variables that are not part of the current dataset, in a time series model that helps to explain or predict the outcome variable in the time series forecasting model, are called future covariates. These external variables are anticipated to influence the predictions in future. These variables help to improve forecasting results in future by adding information that could influence the outcome. For example, in forecasting flight delays, past airport data (historical trends) could be supplemented with local weather data and upcoming major holidays to improve predictions.

Covariates are not necessarily time dependent; they may be time-independent variables too. While weather and holidays are examples of time-dependent covariates, others like gender and weight are time independent.

We discussed future covariates. Let's now peep into the past – past covariates.

In the context of time series forecasting, past covariates are the external variables (outside the dataset) that influenced time series forecasting in the past. To understand the effect of external variables on historical trends, past covariates may be used, for example, climatic conditions or maintenance history that could have influenced the past delays in flights. The difference between past covariates and future covariates, discussed in the previous section, is that the latter are predictors of future outcomes, while the former provide additional context for interpreting forecasts done in the past.

2.4.3.2 Technical Overview of BiTCN

Let's explore the BiTCN, which is an extension of the TCN we discussed earlier:

(a) **Bidirectional Processing**

BiTCN may be understood by extending the TCN model, i.e., by applying convolutions in both forward and backward directions. The first one processes the sequence from start to end (forward) and the other from end to start (backward). Output of the forward pass:

$$y_t^{\text{forward}} = \sum_{k=0}^{K-1} x_{t-k} * w_k^{\text{forward}} + b^{\text{forward}}$$

Output of the backward pass:

$$y_t^{\text{backward}} = \sum_{k=0}^{K-1} x_{T-(t-k)} * w_k^{\text{backward}} + b^{\text{backward}}$$

(b) **Combining Forward and Backward Outputs**

$$y_t = \text{combine}(y_t^{\text{forward}}, y_t^{\text{backward}})$$

The outputs from both directions are combined to produce the final output.

The combined operation is usually done by element-wise addition.

Figure 2-10 shows using three stacked TCN layers to enable conditioning the forecast at $t = t_0 + 1$ on both past and future information. This uses both forward and backward dilated convolutions with kernel size 3 and dilation $2i-1$ for the ith layer.

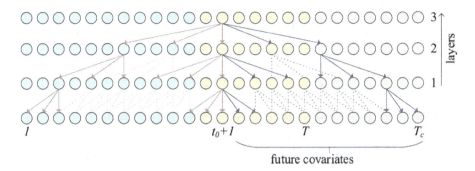

Figure 2-10. *BiTCN architecture [2]*

The blue circles (columns 1-11) represent the input sequence, the yellow circles (columns 12-18) the output sequence, and the green circles (columns 19-25) the additional future covariates on which the forecast can be conditioned.

The red connections (arrows in the middle and left side of the diagram) indicate the backward-looking convolutions, and the purple connections (arrows depicted in the right side of the diagram) are the forward-looking convolutions. For clarity purposes, some inner convolutional connections are shown with dashed lines.

2.4.3.3 BiTCN in Action

Having established a high-level theoretical foundation of BiTCN, we shall now translate abstract concepts into practical code implementation.

Import libraries:

```python
import pandas as pd
import numpy as np
from sklearn.metrics import mean_squared_error, mean_absolute_
percentage_error, r2_score
from IPython.display import display, Markdown
import matplotlib.pyplot as plt
from neuralforecast import NeuralForecast
```

```
from neuralforecast.models import BiTCN
from ray import tune
from neuralforecast.losses.pytorch import GMM, DistributionLoss
from neuralforecast.tsdataset import TimeSeriesDataset
```

Let's load the dataset either from an offline copy or from the neuralforecast.utils dataset, which contains 12 years of monthly air passenger data. Separate the last 12 months of data for training and use the remaining 11 years of data and try to predict.

```
from neuralforecast.utils import AirPassengersDF as Y_df
Y_train_df = Y_df[Y_df.ds<='1959-12-31']
Y_test_df = Y_df[Y_df.ds>'1959-12-31']
dataset, *_ = TimeSeriesDataset.from_df(Y_train_df)
```

Let's initialize and train the BiTCN model by understanding its key parameters.

H is the forecast horizon.

input_size is considered the autorregresive inputs (lags), y=[1,2,3,4] input_size=2 -> lags=[1,2].

Loss is the instantiated train loss class from the losses collection.

max_steps is the maximum number of training steps.

scaler_type is the type of scaler for temporal input normalization; see temporal.

hist_exog_list is the historic exogenous columns.

```
horizon = 12
fcst = NeuralForecast(
    models=[
            BiTCN(h=horizon,
                input_size=12,
                loss=GMM(n_components=7, return_params=True,
                level=[80,90]),
                max_steps=100,
```

```
                scaler_type='standard',
                hist_exog_list=None,

            ),
    ],
    freq='M'
)
fcst.fit(df=Y_train_df)
```

Predict the next defined horizon:

```
y_hat = fcst.predict()
y_hat.set_index('ds',inplace =True)
y_hat.head()
```

BiTCN is the predicted column of interest.

ds	BiTCN	BiTCN-median	BiTCN-lo-90	BiTCN-lo-80	BiTCN-hi-80	BiTCN-hi-90	BiTCN-mu-1	BiTCN-std-1	BiTCN-mu-2	BiTCN-std-2	BiTCN-mu-3	BiTCN-std-3
1960-01-31	412.308014	411.839600	316.411926	336.144257	487.292633	508.321075	392.375031	59.748653	424.555603	41.978580	438.700439	69.711853
1960-02-29	470.638489	474.360596	355.169342	383.567169	553.776306	575.708435	448.301605	58.367268	513.511780	69.798935	458.480072	76.269989
1960-03-31	412.879456	428.226105	244.018051	285.454132	524.314331	553.667786	462.806824	55.989323	393.519043	65.405731	463.916321	34.952991
1960-04-30	401.971649	406.739136	226.265686	261.339966	538.265930	569.001831	318.377930	62.092506	413.927307	39.958599	384.398499	127.513222

Measure the model's accuracy:

```
calculate_error_metrics(Y_test_df[['y']],y_hat[['BiTCN']])
```

```
                MSE : 4996.29248046875
                RMSE : 70.68445587158203
                MAPE : 0.10533150285482407
                r2 : 0.0980561829240788
                adjusted_r2 : 0.007861801216486719
```

Visualize the predictions:

```
Y_train_df.set_index('ds',inplace =True)
```

```
Y_test_df.set_index('ds',inplace =True)
plt.figure(figsize=(20, 3))
y_past = Y_train_df["y"]
y_pred = y_hat[['BiTCN']]
y_test = Y_test_df["y"]
plt.plot(y_past, label="Past time series values")
plt.plot(y_pred, label="Forecast")
plt.plot(y_test, label="Actual time series values")
plt.title('AirPassengers Forecast', fontsize=10)
plt.ylabel('Monthly Passengers', fontsize=10)
plt.xlabel('Timestamp [t]', fontsize=10)
plt.legend();
```

Figure 2-11 helps us to appreciate that the air passenger count predicted by our model is not close to reality. Please note that BiTCN may not always perform better than TCN due to some factors like overfitting, complexity, and some properties of the dataset.

Figure 2-11.   *Actual vs. predicted plot*

2.5 Neural Networks for Sequential Data

In this section, let us discuss neural network architectures that are better suited for sequential data.

2.5.1 Recurrent Neural Network

In this section, let us start our exploration with a type of deep neural network architecture called recurrent neural network (RNN). We will try to understand why it works better with datasets that deal with sequential data like time series data. Sequential data is a type of data where there is a specific order (sequence) in the data. Some examples of sequential data are time series data, speech, audio, and text.

In the RNN architecture, the output from the previous step becomes input for the current step. So, no points for guessing why sequence is important for RNN architecture. This is in contrast to the traditional neural networks where the inputs and outputs from each layer are independent of each other. The emergence of RNN architecture was majorly due to NLP (natural language processing) use cases that required prediction of the next word. RNN provided a solution to this problem by making use of a hidden layer. This layer works as memory, as it is used to remember some information about the sequence.

2.5.1.1 Technical Overview of RNN

In feedforward neural networks, there is only one direction for the data to move from the input layer to the output layer, without any loops. Because of this forward-moving pattern, the data of previous layers will be lost, and no internal memory essentially each input is processed independently. However, in RNN, the data goes through a loop, which means it can remember the past as well as the new data. Information can flow in both directions, with feedback loops that allow the network to maintain a memory of previous inputs.

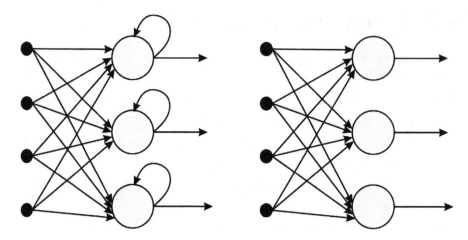

Figure 2-12. *Recurrent neural network and feedforward neural network*

The architecture of RNNs remains the same as any other deep neural network's input and output architecture. The change may be noted in the way information propagates from input to output. Each dense network within a deep neural network has different sets of weights. In contrast, the weight matrices remain the same across the entire recurrent neural network – in other words, same weight matrices across several time steps.

For an input time series $x=\{x1, x2, .., x_n\}$, the RNN computes the hidden state sequence $h = \{h1, h2,.., h_n\}$ as well as the output sequence $y = \{y1; y2;..; y_n\}$ iteratively.

The set of equations used to compute the hidden state sequence and output sequence is

$$h_t = f\left(W_{hx}x_n + W_{hh}h_{n-1} + bh\right)$$

$$y_t = g\left(W_{yh}h_n + b_y\right)$$

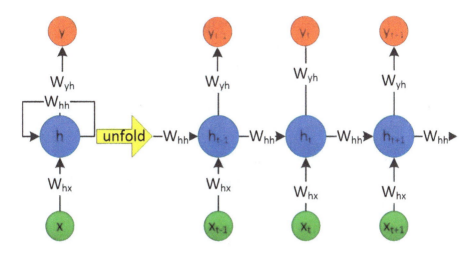

Figure 2-13. *An unrolled RNN*

where

W_{hx} is the input-hidden weight.

W_{hh} is the hidden-hidden weight matrix.

W_{yh} is the hidden-output weight matrix

The input layer **x** takes in the input to the neural network and processes it and passes it to the middle layer. The middle layer **h** can be a stack of multiple hidden layers, each with its own activation functions, weights, and biases, and **y** is the output layer.

The RNN will standardize the different activation functions, weights, and biases so that each hidden layer has the same parameters. Then, instead of creating multiple hidden layers, it will create one and loop over it as many times as required. RNN uses the hidden state h_n at time step n to memorize. The hidden state helps to capture information from the previous steps which helps in better understanding of temporal relationships within data.

Despite the benefits discussed earlier, RNNs have two major challenges: exploding gradient and vanishing gradient. Exploding gradient is a phenomenon that occurs when weights are assigned very large values. Vanishing gradient is a phenomenon that occurs when weights are assigned very small value. This stops the learning process in a neural network. Another issue with RNN is its inability to handle long-range dependencies. Multiple developments have happened to overcome these issues, and one of them is LSTM that we will discuss in the next section.

2.5.1.2 RNN in Action

Having established a high-level theoretical foundation of RNN, we shall now translate abstract concepts into practical code implementation.

Import libraries:

```
import pandas as pd
import numpy as np
from sklearn.metrics import mean_squared_error, mean_absolute_
percentage_error, r2_score
from IPython.display import display, Markdown
import matplotlib.pyplot as plt
from neuralforecast import NeuralForecast
from neuralforecast.losses.pytorch import GMM, MQLoss,
DistributionLoss
from neuralforecast.models import RNN
from neuralforecast.tsdataset import TimeSeriesDataset
from ray import tune
```

Let's load the dataset either from an offline copy or from the **neuralforecast.utils** dataset, which contains 12 years of monthly air passenger count. Separate the last 1 year of data for the test and use the remaining 11 years of data to train the model.

```
from neuralforecast.utils import AirPassengersDF as Y_df

Y_train_df = Y_df[Y_df.ds<='1959-12-31']
Y_test_df = Y_df[Y_df.ds>'1959-12-31']
dataset, *_ = TimeSeriesDataset.from_df(Y_train_df)
```

Let's initialize and train the model (RNN) by understanding its key parameters.

H is the forecast horizon.

input_size is the maximum sequence length for truncated train backpropagation. Default –1 uses all history.

inference_input_size is the maximum sequence length for truncated inference. Default –1 uses all history.

Loss is the instantiated train loss class from the losses collection.

scaler_type is the step size between each window of temporal data.

encoder_n_layers is the number of layers for the RNN.

encoder_hidden_size is the unit for the RNN's hidden state size.

context_size is the size of context vector for each timestamp on the forecasting window.

decoder_hidden_size is the size of the hidden layer for the MLP decoder.

decoder_layers is the number of layers for the MLP decoder.

max_steps is the maximum number of training steps.

```
horizon = 12
fcst = NeuralForecast(
    models=[RNN(h=horizon,
                input_size=-1,
                inference_input_size=24,
                loss=MQLoss(level=[80, 90]),
                scaler_type='robust',
                encoder_n_layers=2,
                encoder_hidden_size=128,
                context_size=10,
                decoder_hidden_size=128,
                decoder_layers=2,
                max_steps=300,
                #futr_exog_list=['y_[lag12]'],
```

```
                #hist_exog_list=['y_[lag12]'],
                #stat_exog_list=['airline1'],
                )
    ],
    freq='M'
)

fcst.fit(df=Y_train_df, val_size=12)
```

Predict for the next defined horizon.

```
y_hat = fcst.predict()
y_hat.set_index('ds',inplace =True)
y_hat.head()
```

RNN-median is the predicted column of interest.

ds	RNN-median	RNN-lo-90	RNN-lo-80	RNN-hi-80	RNN-hi-90
1960-01-31	371.641907	337.374939	348.417114	384.715179	387.867615
1960-02-29	371.430481	341.607483	341.278748	384.412750	389.489380
1960-03-31	379.759338	349.267303	353.357025	393.845795	402.789429
1960-04-30	382.029968	358.380554	362.446228	397.847046	408.122864
1960-05-31	385.985779	369.793579	372.185944	409.069183	419.553223
1960-06-30	437.951172	424.153687	426.440796	459.305328	473.591187

Measure the model's accuracy:

```
calculate_error_metrics(Y_test_df[['y']],y_hat[['RNN-median']])
```

```
            MSE : 4642.16552734375
            RMSE : 68.13343811035156
            MAPE : 0.12443191558122635
            r2 : 0.1619841548291906
            adjusted_r2 : 0.07818257031210973
```

Visualize the predictions:

```
Y_train_df.set_index('ds',inplace =True)
Y_test_df.set_index('ds',inplace =True)
plt.figure(figsize=(20, 3))
y_past = Y_train_df["y"]
y_pred = y_hat[['RNN-median']]
y_test = Y_test_df["y"
plt.plot(y_past, label="Past time series values")
plt.plot(y_pred, label="Forecast")
plt.plot(y_test, label="Actual time series values")
plt.title('AirPassengers Forecast', fontsize=10)
plt.ylabel('Monthly Passengers', fontsize=10)
plt.xlabel('Timestamp [t]', fontsize=10)
plt.legend();
```

Figure 2-14 helps us to appreciate that the air passenger count predicted by our model is close to reality.

Figure 2-14. *Actual vs. predicted plot*

2.5.2 Long Short-Term Memory

The long short-term memory (LSTM) architecture is a modification to the RNN architecture to allow additional signal paths. These additional paths help in bypassing many processing steps encountered at each stage of the network. This modification helps in remembering information over a large number of time steps. While this modification improves the performance compared to the RNN, it also introduces additional complexity. This complexity has an impact on training speed compared to RNNs. Popular tools like Apple's Siri and Google's AlphaGo were based on LSTM.

2.5.2.1 Technical Overview of LSTM

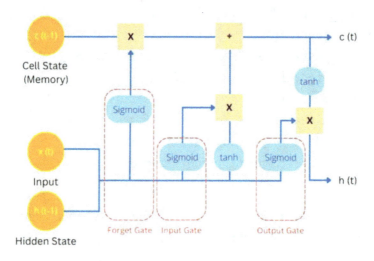

Figure 2-15. *Architecture of LSTM*

From Figure 2-15, it is evident that for the input for each computational step, three values are considered. The values are (a) current input x(t), (b) past value of hidden state h(t-1), and (c) past value of short-term memory c(t-1). Next, these inputs pass through three gates – forget gate, input gate, and output gate – before obtaining a new cell state c(t) and hidden state h(t). Let us understand these three gates in some more detail.

(a) **Forget Gate**

In this gate, a decision is taken with regard to which current and previous information is retained and discarded. The decision is taken on past values of the hidden state and values of the current input. These values are passed through a sigmoid function.

58

Those who are familiar with logistic regression would recall that the sigmoid function's output ranges between 0 and 1. In our context, the value 0 means that previous information can be discarded. This is due to possible availability of new, more important information. The value 1 means that the previous information is preserved. The resultant output of the sigmoid function is multiplied by the current cell state so that knowledge that is no longer needed is discarded since it is multiplied by 0.

(b) **Input Gate**

In this gate, a decision is taken to evaluate the current input to solve the task. To achieve this, the current input is multiplied by the hidden state and the weight matrix of the previous run. All the information that appears important in the input gate is then added to the cell state. The resultant forms the new cell state c(t). This new cell state becomes the current state of the long-term memory to be used in the subsequent run.

(c) **Output Gate**

In this gate, the output of the LSTM model is calculated for the hidden state. Depending on the application, it can be, for example, a word that complements the meaning of the sentence. In order to compute h(t), the sigmoid function is used to decide what information can pass through the output gate. The result is then multiplied by cell state, after c(t) passes through the tanh activation function.

Despite the modifications, all RNN-based architectures come with an inherent limitation – they do not support parallel processing. You would have noted by now that propagation paths in RNNs increase linearly with the number of steps in sequence. It is not possible to leverage powerful parallel processing capable processors like GPUs and TPUs (within a single training example) due to the sequential processing architecture in RNNs. In the next chapter, we explore a newer architecture called transformer that overcomes the limitations of the RNN architecture.

2.5.2.2 LSTM in action

Having established a high-level theoretical foundation of LSTM, we shall now translate abstract concepts into practical code implementation.

Import libraries:

```
import pandas as pd
import numpy as np
from sklearn.metrics import mean_squared_error, mean_absolute_
percentage_error, r2_score
from IPython.display import display, Markdown
import matplotlib.pyplot as plt
from neuralforecast import NeuralForecast
from neuralforecast.models import LSTM
from neuralforecast.tsdataset import TimeSeriesDataset
from neuralforecast.losses.pytorch import GMM, MQLoss,
DistributionLoss
from neuralforecast.utils import AirPassengersDF as Y_df
from ray import tune
```

Let's load the dataset either from an offline copy or from the **neuralforecast.utils** dataset, which contains 12 years of monthly air passenger count. Separate the last 1 year of data for the test and use the remaining 11 years of data to train the model.

```
from neuralforecast.utils import AirPassengersDF as Y_df

Y_train_df = Y_df[Y_df.ds<='1959-12-31'] # 132 train data
Y_test_df = Y_df[Y_df.ds>'1959-12-31']   # 12 test data
```

Let's initialize and train the LSTM model by understanding its key parameters.

H is the forecast horizon.

input_size is the maximum sequence length for truncated train backpropagation. Default –1 uses all history.

Loss is the instantiated train loss class from the losses collection.

scaler_type is the step size between each window of temporal data.

encoder_n_layers is the number of layers for the RNN.

encoder_hidden_size is the unit for the RNN's hidden state size.

context_size is the size of the context vector for each timestamp on the forecasting window.

decoder_hidden_size is the size of the hidden layer for the MLP decoder.

decoder_layers is the number of layers for the MLP decoder.

max_steps is the maximum number of training steps.

```
horizon = 12
fcst = NeuralForecast(
    models=[LSTM(h=horizon, input_size=-1,
                loss=DistributionLoss(distribution='Normal',
                level=[80, 90]),
                scaler_type='robust',
                encoder_n_layers=2,
                encoder_hidden_size=128,
                context_size=10,
                decoder_hidden_size=128,
```

```
                    decoder_layers=2,
                    max_steps=200,

                    )
    ],
    freq='M'
)
fcst.fit(df =Y_train_df)

model.fit(dataset=dataset)
```

Predict the next defined horizon:

```
y_hat = fcst.predict()
y_hat.set_index('ds',inplace =True)
y_hat.head()
```

LSTM is the predicted column of interest.

	LSTM	LSTM-median	LSTM-lo-90	LSTM-lo-80	LSTM-hi-80	LSTM-hi-90
ds						
1960-01-31	350.888092	350.775146	295.248077	305.006714	396.771301	410.420013
1960-02-29	363.306000	363.748505	323.620117	330.534210	396.106293	407.590698
1960-03-31	390.778564	390.904602	339.590057	350.849121	432.370148	445.618408
1960-04-30	424.257782	423.604706	372.827545	383.003418	464.217773	476.210663
1960-05-31	449.194122	450.273926	397.561066	412.049072	484.153259	494.906708
1960-06-30	473.402252	473.729187	415.734924	430.352264	517.406128	530.791321

Measure the model's accuracy:

```
calculate_error_metrics(Y_test_df[['y']],y_hat[['LSTM']])
```

```
MSE : 2924.734375
RMSE : 54.080814361572266
MAPE : 0.10070467740297318
r2 : 0.47201929308164103
adjusted_r2 : 0.4192212223898052
```

Visualize the predictions:

```
Y_train_df.set_index('ds',inplace =True)
Y_test_df.set_index('ds',inplace =True)
plt.figure(figsize=(20, 3))
y_past = Y_train_df["y"]
y_pred = y_hat[['LSTM']]
y_test = Y_test_df["y"]
plt.plot(y_past, label="Past time series values")
plt.plot(y_pred, label="Forecast")
plt.plot(y_test, label="Actual time series values")
plt.title('AirPassengers Forecast', fontsize=10)
plt.ylabel('Monthly Passengers', fontsize=10)
plt.xlabel('Timestamp [t]', fontsize=10)
plt.legend();
```

Figure 2-16 helps us to appreciate that the air passenger count predicted by our model is close to reality.

Figure 2-16. *Actual vs. predicted plot*

2.6 Neural Networks Based on Autoregression

In this section, let us discuss neural networks that leverage autoregression (AR).

We are going to cover the DeepAR forecasting method that is based on autoregressive recurrent networks. Autoregression models in time series forecast future values by depending on past observations of the same variable. With a fundamental assumption that past and future values of the same variable are dependent, they use a linear combination of past observations for time series forecasts. The value of "order" of the model is nothing but the number of past values used in computing the future value.

The techniques we have covered so far provided us with a single predicted forecast value. However, the technique we are going to cover in this section is a probabilistic forecasting technique. Probabilistic forecasting techniques have a unique feature. These classes of techniques do not forecast a single value; rather, they provide a range of values that we call, in the language of probability, probability distribution. Acknowledging that the future is inherently uncertain, the output of probabilistic forecasting is a range of possible outcomes of a forecasted variable.

There are many examples of probabilistic forecasting that we encounter in real life; weather forecasts are a classic example. For instance, the temperature forecast that I saw last week in New Jersey was displayed as follows: 70% chance of heat on Monday would be between 81 and 83 F, 60% chance of heat on Tuesday between 76 and 79 F.

2.6.1 Key Features of Probabilistic Forecasting

a) **Quantification of uncertainty**: A spectrum of value is provided with probabilities instead of a single forecasted value. This helps in quantifying uncertainty associated with the future.

b) **Better decisions with risk considerations**: The inherent uncertainty in future time periods, which is important in taking strategic decisions that take into account the risks involved, is supported by the range and likelihood of outcomes.

c) **Percentile representation**: The forecasts can be expressed in percentiles and represented using box plots and whisker plots representing different confidence levels.

The early traditional forecasting methods were developed in the context of time series forecasting individual time series data. The scope later expanded to forecast a small number of groups of time series. In the early traditional approaches, model parameters for each given time series within the group were independently estimated from historical observations. The model was then manually selected to cater to various parameters like trend, seasonality, cycles, and autocorrelation. The best fit model was then used in time series forecasting as per the model dynamics. DeepAR is good at handling complex time series with seasonality, trends, and other irregularities.

In the last decade, we have seen an explosion in data availability. New tools and techniques to handle big data became popular. New expectations and associated developments to handle use cases that demanded forecasting in the order of millions of related time series emerged. Let us appreciate this scenario by citing a few use cases. Forecasting energy demands of large apartment complexes, forecasting power consumption of server farms, and forecasting demand for individual products during Thanksgiving sales are a few examples.

The common aspect you may have observed in these scenarios is the availability of large amount of historical data. This data could be of same or similar events. The time series data from similar events can be utilized in time series forecasting for individual time series. There are two major advantages of using time series data from similar events; they are (a) efficiently fitting more complex models and (b) reducing the effort in feature engineering and model selection steps. The DeepAR model for time series forecasting efficiently learns from historical data leveraging these two advantages.

The DeepAR model uses recurrent neural networks (RNNs) to learn temporal dependencies and patterns in the data. The model takes past values of a variable and generates a probability distribution of future values. This distribution can be used to estimate the most likely future values or to generate confidence intervals for predictions.

Despite the advantages discussed earlier, regarding the usage of learning from multiple time series, a few practical problems exist. In real-world datasets, the magnitude of time series varies widely. You will also note that the distribution of magnitudes is strongly skewed. As an example, we can see the plot in Figure 2-17, which explains the distribution of sales velocity of items sold (in millions) by a leading online retailer. Sales velocity is defined as the average weekly sales of a product.

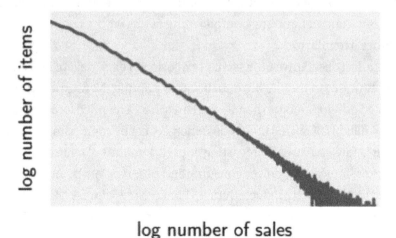

Figure 2-17. *Log-log histogram of the number of items versus number of sales [7]*

A few approaches were suggested based on group-based regularization techniques, which largely became inapplicable because of variations in velocities within individual groups. Also, skewed distributions limit the use of normalization methods like input standardization or batch normalization.

2.6.2 Technical Overview of Deep Autoregressive

The DeepAR model has some key benefits compared to traditional approaches. The major advantages that set DeepAR apart are as follows: (a) Relatively much less effort and time need to be spent on feature engineering to capture complex and group-dependent behavior. This is because the model learns seasonality and dependencies on given covariates across the time series. (b) It has an ability to provide forecasts for products with little to no historical data. This is because of learning from historical data of similar events.

The DeepAR model has properties that help produce better forecasts by learning from historical behavior of all the time series taken together. (a) It incorporates a wide range of likelihood functions. This allows the time series modeling team to choose suitable functions based on statistical properties of the data. (b) The probabilistic forecasts are generated in the form of Monte Carlo Samples. These can be used to compute quantile estimates belonging to subranges in the prediction horizon. This is important as discussed earlier in this section, where we pointed out advantages of forecasts with probabilities compared to a point forecast value.

In Figure 2-18, we see two parts – to the left is the input and to the right is the prediction.

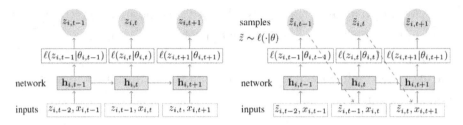

Figure 2-18. *Summary of the model [7]*

The input to the network consists of three parameters: (a) covariates $x_{i,t}$, (b) preceding time step's target value $z_{i,t-1}$, and (c) preceding network output value $h_{i,t-1}$.

The output of the network $h_{i,t}$ is then used to compute the parameters of the likelihood. These parameters in turn are used to train the model parameters. In order to perform prediction, the history of the time series $z_{i,t}$ is provided as input for t < t0, then in the prediction range (right) for t ≥ t0 a sample is drawn and fed to the next point. This process continues till the end of the prediction range t = t0 + T. This results in the generation of a single sample trace. Multiple traces representing the joint predicted distribution are generated by repeating the steps for prediction.

To understand the mathematical model of DeepAR, we need to remember that the primary goal is to model conditional distribution, where $Z_{i,t}$ is the value of time series i at an instance of time t.

$$P\left(\mathbf{z}_{i,t_0:T} \middle| \mathbf{z}_{i,1:t_0-1}, \mathbf{x}_{i,1:T}\right)$$

This conditional distribution is for the future of each time series in $\left[z_{i,t_0}, z_{i,t_0+1}, \ldots, z_{i,T}\right] \quad := \quad \mathbf{z}_{i,t_0:T}.$

The past values are
$\left[z_{i,1}, \ldots, z_{i,t_0-2}, z_{i,t_0-1}\right] := \mathbf{z}_{i,1:t_0-1}.$

Here, t_0 denotes the time point from which we assume $Z_{i, t}$ to be unknown at prediction time. The $Z_i, 1{:}t_0{-}1$ and $X_{i, 1:T}$ are covariates that are assumed to be known for all time points. To avoid confusion in terminology, we skip using terms like "past" or "future"; rather, we use time ranges $[1, t_0 {-}1]$ and $[t_0, T]$ as conditioning range and prediction range. While training the model, both time ranges have to be in the past so that $Z_{i,t}$ is observed. However, during prediction $Zi_{,t}$ is available in the conditioning range only.

Note   The time index t is relative, i.e., t = 1 can correspond to a different actual time period for each i.

2.6.3 DeepAR in Action

Having established a high-level theoretical foundation of DeepAR, we shall now translate abstract concepts into practical code implementation.

Import libraries:

```
import pandas as pd
import numpy as np
from sklearn.metrics import mean_squared_error, mean_absolute_
percentage_error, r2_score
from IPython.display import display, Markdown
import matplotlib.pyplot as plt
from neuralforecast import NeuralForecast
from neuralforecast.losses.pytorch import MQLoss,
DistributionLoss, GMM, PMM
from neuralforecast.tsdataset import TimeSeriesDataset
import pandas as pd
import pytorch_lightning as pl
import matplotlib.pyplot as plt
from neuralforecast.models import DeepAR
```

```
from neuralforecast.losses.pytorch import DistributionLoss,
HuberMQLoss
from neuralforecast.utils import AirPassengers,
AirPassengersPanel, AirPassengersStatic
from neuralforecast.utils import AirPassengers,
AirPassengersPanel, AirPassengersStatic, AirPassengersPanel,
AirPassengersStatic
AirPassengersPanel.head()
```

	unique_id	ds	y	trend	y_[lag12]
0	Airline1	1949-01-31	112.0	0	112.0
1	Airline1	1949-02-28	118.0	1	118.0
2	Airline1	1949-03-31	132.0	2	132.0
3	Airline1	1949-04-30	129.0	3	129.0
4	Airline1	1949-05-31	121.0	4	121.0

```
print(AirPassengersStatic)
```

	unique_id	airline1	airline2
0	Airline1	0	1
1	Airline2	1	0

Let's load the dataset either from an offline copy or from the **neuralforecast.utils** dataset, which contains 12 years of monthly air passenger count. Separate the last 1 year of data for the test and use the remaining 11 years of data to train the model.

```
Y_train_df = AirPassengersPanel[AirPassengersPanel.ds<AirPassen
gersPanel['ds'].values[-12]]
Y_test_df = AirPassengersPanel[AirPassengersPanel.ds>=AirPassen
gersPanel['ds'].values[-12]].reset_index(drop=True)
```

Let's initialize and train the DeepAR model by understanding its key parameters.

h is the forecast horizon.

input_size is the autoregressive input size, y=[1,2,3,4] input_size=2 -> y_[t-2:t]=[1,2].

lstm_n_layers is the number of LSTM layers.

trajectory_samples is the number of Monte Carlo trajectories during inference.

Loss is the instantiated train loss class from the losses collection.

learning_rate is the learning rate between (0, 1).

stat_exog_list is the static exogenous columns.

futr_exog_list is the future exogenous columns.

max_steps is the maximum number of training steps.

val_check_steps is the number of training steps between every validation loss check.

early_stop_patience_steps is the number of validation iterations before early stopping.

scaler_type is the type of scaler for temporal input normalization.

```
nf = NeuralForecast(
    models=[DeepAR(h=12,
                   input_size=48,
                   lstm_n_layers=3,
                   trajectory_samples=100,
                   loss=DistributionLoss(distribution='Normal',
                   level=[80, 90], return_params=False),
                   learning_rate=0.005,
```

```
                    stat_exog_list=['airline1'],
                    futr_exog_list=['trend'],
                    max_steps=100,
                    val_check_steps=10,
                    early_stop_patience_steps=-1,
                    scaler_type='standard',
                    enable_progress_bar=True),
    ],
    freq='M'
)
nf.fit(df=Y_train_df, static_df=AirPassengersStatic, val_
size=12)
```

Predict the next defined horizon:

```
Y_hat_df = nf.predict(futr_df=Y_test_df)
Y_hat_df.head()
```

DeepAR is the predicted column of interest.

unique_id	ds	DeepAR	DeepAR-median	DeepAR-lo-90	DeepAR-lo-80	DeepAR-hi-80	DeepAR-hi-90
Airline1	1960-01-31	415.200775	413.846069	385.291718	394.683624	443.522583	454.160156
Airline1	1960-02-29	431.533722	431.639160	394.706299	401.953308	460.426178	467.867493
Airline1	1960-03-31	437.992218	439.715088	395.064148	407.487457	467.977692	476.772095
Airline1	1960-04-30	446.789795	447.080872	406.911774	415.343719	473.718048	484.828094
Airline1	1960-05-31	464.116455	466.025696	424.759277	433.739288	491.210815	497.137115
Airline1	1960-06-30	517.927979	520.244934	478.387085	489.869629	543.724243	551.895020

Measure the model's accuracy:

```
calculate_error_metrics(Y_test_df[['y']],Y_hat_df[['DeepAR']])
```

```
MSE : 455.6470642089844
RMSE : 21.34589195251465
MAPE : 0.02916746586561203
r2 : 0.9837497995847083
adjusted_r2 : 0.9830111541112859
```

Visualize the predictions:

```
Y_hat_df =Y_hat_df.reset_index(drop=False).drop(columns=['uniq
ue_id','ds'])
plot_df = pd.concat([Y_test_df, Y_hat_df], axis=1)
plot_df = pd.concat([Y_train_df, plot_df])
plt.figure(figsize=(20, 3))
plot_df = plot_df[plot_df.unique_id=='Airline1'].drop('unique_
id', axis=1)
plt.plot(plot_df['ds'], plot_df['y'], c='black', label='True')
plt.plot(plot_df['ds'], plot_df['DeepAR-median'], c='blue',
label='median')
plt.fill_between(x=plot_df['ds'][-12:],
                 y1=plot_df['DeepAR-lo-90'][-12:].values,
                 y2=plot_df['DeepAR-hi-90'][-12:].values,
                 alpha=0.4, label='level 90')
plt.title('AirPassengers Forecast', fontsize=10)
plt.ylabel('Monthly Passengers', fontsize=10)
plt.xlabel('Timestamp [t]', fontsize=10)
plt.legend()
plt.grid()
plt.plot()
```

Figure 2-19 helps us to appreciate that the air passenger count predicted by our model is close to reality.

Figure 2-19. *Actual vs. predicted plot*

2.7 Neural Basis Expansion Analysis

In this section, we cover neural basis expansion analysis for time series (NBEATS). This is an effective but simple architecture; let us learn how. This architecture is built with a deep stack of MLPs with the doubly residual connections. Depending on the blocks used, NBEATS has a generic and interpretable architecture. In use cases involving scarce data settings, the interpretable architecture of NBEATS is recommended. The primary reason is that the model regularizes its predictions by expressing in terms of constituent harmonics and trends. This makes it a suitable choice for many forecasting tasks.

2.7.1 Technical Overview of NBEATS

NBEATS architecture was developed in the process of exploring the use of deep learning to solve univariate time series forecasting use cases. This architecture is designed using a deep learning network with multiple fully connected layers and relies on a network of backward and forward residual links. The intention of including this model in our journey to understand time series forecasting with GenAI is that this model was the first pure deep learning approach that performed better than existing statistical approaches in the Makridakis M-competition. The NBEATS model surpassed the winning solution of the M4 competition. You may want to look at https://forecasters.org/resources/time-series-data/ to know more and participate.

The building blocks of the architecture are "stacks." NBEATS consists of a layer of stacks. The individual stack is used to focus on various levels of temporal resolution. For example, one stack may be used to focus on long-term trends, while another stack may be used to focus on the short-term seasonality component. Each stack also has a series of "blocks." These blocks are responsible for capturing a specific temporal pattern like trend or seasonality. The blocks have backcast and forecast components. These help to learn from past behavior and help in time series forecasting based on the patterns learned.

The NBEATS architecture has advantages like interpretability (a challenge with neural networks in general), faster to train, and applicability to a wide spectrum of use cases in many domains with minimal to no changes in architecture. This is achieved with the help of generic architecture and interpretable architecture, which are discussed later.

From Figure 2-20, it is evident that the NBEATS architecture is a multilayered fully connected (FC) network. This network also has ReLU nonlinearities. A fully connected layer in a neural network means that in the neural network, each input node is mapped to an output node. This is in contrast to a convolutional layer, where you will find unconnected nodes. Going back to the architecture diagram, the predictions include forward basis expansion coefficient, forecast θ^f, and backward basis expansion coefficient, backcast θ^b. Using doubly residual stacking principle, the blocks are organized into stacks. The stack includes layers with shared forecasts and backcasts. Developing a deep learning network with interpretable forecasts is possible by hierarchical aggregation (adding) of the forecasts.

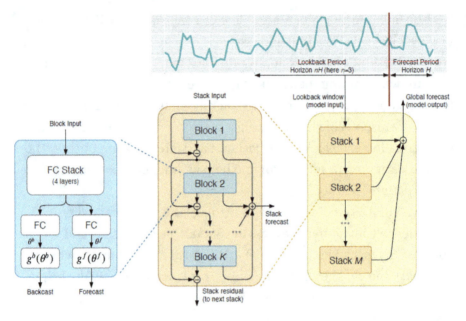

Figure 2-20. *Architecture of NBEATS [3]*

The residual stacking principle is a concept where each block iteratively updates the forecast by considering the residual error from the preceding block.

The forward basis expansion coefficient, forecast θ^f, is accumulated to generate the final prediction. The backward basis expansion coefficient, backcast θ^b, is used to adjust the input series iteratively. The input to each block is the residual time series. Residual time series is the remaining value (residual) after outputs from the preceding block have been subtracted. The input traverses through multiple FC layers.

There are two configurations of the NBEATS architecture. (a) Generic architecture: The generic architecture does not rely on any time series–specific (trend, seasonality) knowledge. The model learns the TS patterns directly from the dataset. (b) Interpretable architecture: The interpretable architecture is built by modifying the architecture shown in Figure 2-13. The architecture has a trend block and seasonality block. These blocks help to capture the trend and seasonality in the dataset.

2.7.2 NBEATS in Action

Having established a high-level theoretical foundation of NBEATS, we shall now translate abstract concepts into practical code implementation.

Import libraries:

```python
import pandas as pd
import numpy as np
from sklearn.metrics import mean_squared_error, mean_absolute_
percentage_error, r2_score
from IPython.display import display, Markdown
import matplotlib.pyplot as plt
from ray import tune
from neuralforecast.core import NeuralForecast
from neuralforecast.models import NBEATS, NHITS
```

Let's load the dataset either from an offline copy or from the **neuralforecast.utils** dataset, which contains 12 years of monthly air passenger count. Separate the last 1 year of data for the test and use the remaining 11 years of data to train the model.

```python
from neuralforecast.utils import AirPassengersDF
Y_df = AirPassengersDF
Y_df = Y_df.reset_index(drop=True)
Y_df.head()
```

	unique_id	ds	y
0	1.0	1949-01-31	112.0
1	1.0	1949-02-28	118.0
2	1.0	1949-03-31	132.0
3	1.0	1949-04-30	129.0
4	1.0	1949-05-31	121.0

```
train_data = Y_df.head(132)
test_data = Y_df.tail(12)
```

Let's initialize and train the NBEATS model by understanding its key parameters.

h is the forecast horizon.

input_size is considered the autoregressive inputs (lags), y=[1,2,3,4] input_size=2 -> lags=[1,2].

max_steps is the maximum number of training steps.

```
horizon = 12
models = [NBEATS(input_size=2 * horizon, h=horizon, max_
steps=50)]
nf = NeuralForecast(models=models, freq='M')
nf.fit(df=train_data)
```

Predict the next defined horizon:

```
Y_hat_df = nf.predict().reset_index()
Y_hat_df.head()
```

NBEATS is the predicted column of interest.

	unique_id	ds	NBEATS
0	1.0	1960-01-31	427.045990
1	1.0	1960-02-29	448.289337
2	1.0	1960-03-31	453.882721
3	1.0	1960-04-30	455.047668
4	1.0	1960-05-31	496.515900

Measure the model's accuracy:

```
calculate_error_metrics(test_data[['y']],Y_hat_df['NBEATS'])
```

```
MSE : 1111.5928955078125
RMSE : 33.34056091308594
MAPE : 0.06422501802444458
r2 : 0.799332357382532
adjusted_r2 : 0.7792655931207852
```

Visualize the predictions:

```
train_data.set_index('ds',inplace =True)
Y_hat_df.set_index('ds',inplace =True)
test_data.set_index('ds',inplace =True)
plt.figure(figsize=(20, 3))
item_id = "airline_1"
y_past = train_data["y"]
y_pred = Y_hat_df['NBEATS']
y_test = test_data["y"]
plt.plot(y_past, label="Past time series values")
plt.plot(y_pred, label="Mean forecast")
plt.plot(y_test, label="Actual time series values")
plt.title('AirPassengers Forecast', fontsize=10)
plt.ylabel('Monthly Passengers', fontsize=10)
plt.xlabel('Timestamp [t]', fontsize=10)
plt.legend();
```

Figure 2-21 helps us to appreciate that the air passenger count predicted by our model is close to reality.

Figure 2-21. *Actual vs. predicted plot*

2.8 Summary

In this chapter, we discussed various neural network architectures that are used for time series forecasting. We covered architectures based on CNN, RNN, and LSTM that can be leveraged for time series forecasting. We understood terms like dilated convolutions, causal convolutions, future covariates, and mathematical overview of some of the models. We discussed temporal convolution networks and how they are useful in handling sequential data.

We understood how DeepAR works and its effectiveness in handling complex time series datasets. We explored how an effective yet simple architecture like NBEATS works. Finally, we saw the models in action by implementing them with use cases for prediction.

Please use a dataset of your choice to practice. In the next chapter, we will move a step closer to GenAI by understanding the transformer architecture and how it helps in time series forecasting. Transformers are building blocks for training GenAI models.

2.9 References

[1].   GitHub – autogluon/autogluon: Fast and Accurate ML in 3 Lines of Code

[2].   Olivier Sprangers, Sebastian Schelter, Maarten de Rijke (2023). Parameter-Efficient Deep Probabilistic Forecasting. **https://doi.org/10.1016/j.ijforecast.2021.11.011**

[3].   Boris N. Oreshkin, Dmitri Carpov, Nicolas Chapados, Yoshua Bengio (2019). "N-BEATS: Neural basis expansion analysis for interpretable time series forecasting." https://doi.org/10.48550/arXiv.1905.10437

[4].   WaveNet: A generative model for raw audio. Computing
        Research Repository. https://doi.org/10.48550/arXiv
        .1609.03499

[5].   Exponential Pixelating Integral transform with
        dual fractal features for enhanced chest X-ray
        abnormality detection. https://doi.org/10.1016/j.
        compbiomed.2024.109093

[6].   The predictive skill of convolutional neural networks
        models for disease forecasting. https://doi.
        org/10.1371/journal.pone.0254319.g003

[7].   DeepAR: Probabilistic Forecasting with Autoregressive
        Recurrent Networks. https://doi.org/10.48550/
        arXiv.1704.04110

CHAPTER 3

Transformers for Time Series

Chapter Goal: Learn how to leverage the different types of transformers and solve time series problems.

In the preceding chapter, we explored different kinds of neural network architectures and practically implemented them using real-world datasets.

This chapter focuses on breaking down transformers, understanding them at a high level, and exploring other popular transformer variants. Let us also understand how they can be used to solve time series problems.

3 Introduction to Transformers

Transformers initially revolutionized natural language processing and have increasingly found their application in other realms such as computer vision, audio processing, bioinformatics, finance, robotics, and time series analysis. This chapter delves into the core concepts of transformers and explains how these powerful models can be adapted to effectively handle time series data.

© Banglore Vijay Kumar Vishwas and Sri Ram Macharla 2025
B. V. Vishwas and Sri Ram Macharla, *Time Series Forecasting Using Generative AI*,
https://doi.org/10.1007/979-8-8688-1276-7_3

We will dissect the transformer's architecture by breaking it down into its fundamental components and understand how it works internally. Beyond the foundation of the transformer, we will explore a diverse range of variants that have been specifically tailored for time series analysis. These innovative architectures offer different advantages in handling various time series challenges.

By the end of this chapter, readers will have a solid grasp of transformers and their potential in the time series domain, enabling them to effectively apply to solve their own problems.

"Attention Is All You Need" [1] was the paper that introduced the transformer architecture; this revolutionized the natural language processing by demonstrating the power of the attention mechanism. Numerous efforts have tried to push the boundaries of recurrent language models and encoder-decoder architectures. Sequence learning architectures, such as gated recurrent neural networks, recurrent neural networks, and long short-term memory in particular, have been firmly established as state-of-the-art approaches in sequence modeling any data that exhibits a sequential pattern.

3.1 Technical Overview of Transformers

Let's break down the components that underpin the transformer's remarkable performance and how it works through the lenses of the original paper [1].

Transformers are designed using stacked self-attention and point-wise, fully connected layers for both the encoder and decoder, as shown in Figure 3-1.

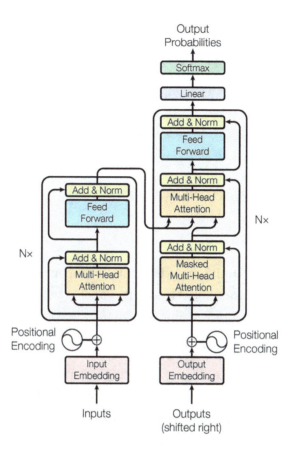

Figure 3-1. *Transformer model architecture [1]*

Transformers are built on encoder-decoder architecture. The encoder applies the mathematical function to the data and transforms input to a certain representation, while the decoder applies the inverse function to recover back the original data.

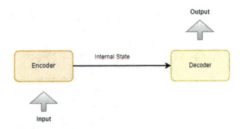

Figure 3-2. *Encoder-decoder*

Let's understand the components from the left (bottom-up approach) of Figure 3-1:

a) **Input Embedding**

   This is the initial step where raw text is converted into the format suitable for the model to process which is numerical representation.

b) **Positional Encoding**

   Positional encoding is a method used in transformers to incorporate word order by assigning a unique number to each word in a sentence, allowing the model to learn sequence information. This is a way of saving the word order in the data itself rather than the network.

   As the original paper had no recurrent and no convolution for the model to make use of the order of the sequence, there was a need to inject some information about the relative or absolute position of the tokens in the sequence. The positional encoding has the same dimensions d_{model} as the embedding so that they can be summed up:

$$PE_{(pos,2i)} = sin(pos/10000^{2i/d_{model}})$$
$$PE_{(pos,2i+1)} = cos(pos/10000^{2i/d_{model}})$$

where d_{model} is the embedding dimension, positional encoding is a matrix *PE* of shape (n, d_{model}), *pos* is the position, and *i* is the dimension where each dimension of the positional encoding corresponds to a sinusoid. Sine and cosine functions of different frequencies are used in the example; however, we can explore other encoding techniques such as rotary positional encoding, no positional encoding, and absolute positional encoding.

c) **Encoder**

Let's dive deep into the encoder part of the transformers, which is highlighted below; it consists of a stack of six identical layers. Each layer consists of a multi-head self-attention mechanism and a point-wise fully connected feedforward network. There is a residual connection around each of the two sublayers followed by layer normalization.

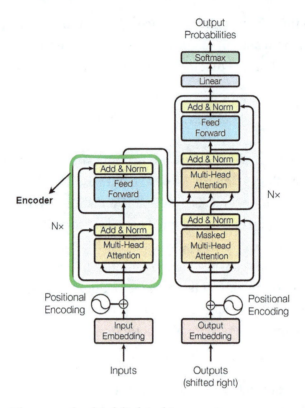

Figure 3-3. *The encoder highlighted in green*

d) **Attention**

An attention function can be described as mapping a query and a set of key-value pairs to an output, where the query, keys, values, and output are all vectors. The output is computed as a weighted sum.

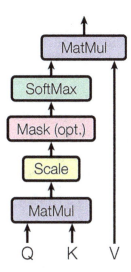

Figure 3-4. *Scaled dot-product attention*

Scaled dot-product attention is used to enable models to capture complex dependencies between input elements. The input consists of **query** (Q) that represents what you are looking for, **key** (K) that represents what you are searching with, and **value** (V) that represents information you want to retrieve.

Dot product is used to calculate the similarity between the query and each key. A higher result means a closer match. **Scaling** is used to prevent dot products from getting too large; as the dimension of keys increases, the dot product tends to grow large. A large value in the softmax function can lead to gradients that are close to zero, slowing down training as the model parameters receive negligible updates and slowing down the learning; this is called the vanishing gradient problem. **Weighted**

sum multiplies each value by the corresponding attention weight and sums them up to get the final output.

Scaled dot-product attention helps the model to focus on the most relevant parts of the data passed when generating the output. Let us understand scaled dot-product attention with a simple example.

(Q): Imagine you have a question about time series forecasting.

(K): Scan for keywords in the text.

(V): Find relevant keywords like trend, seasonality, forecasting, etc.

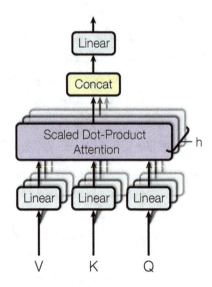

Figure 3-5. *Depiction of multi-head attention*

Multi-head Attention

Multi-head attention is multiple attention working in parallel, which allows transformer models to focus on different parts of the input sequence simultaneously. Multi-head attention allows the model to jointly attend to information from different representation subspaces at different positions.

Multiple attention is projected instead of a single attention function with dmodel – dimensional keys, value, and queries. K, Q, and V are linearly projected h times with different linear projections dk, dq, and dv dimensions, respectively. The model jointly sees the information from different representation subspaces at different positions.

e) **Feedforward Neural Network**

This layer applies the same feedforward neural network to each position separately and identically. It consists of two linear transformations with a ReLU activation in between.

$$\text{FFN}(x) = \max(0, xW_1 + b_1)W_2 + b_2$$

f) **Layer Normalization**

This is applied after each sublayer to stabilize or improve the performance of the deep neural network.

g) **Residual Connections**

This helps in mitigating the vanishing gradient problem, which is essentially when the gradients, used to update the weights of the network during training, become extremely small as they propagate backward through the layers and also act as a shortcut path to bypass one or more layers.

Let's now understand the components from the right (bottom-up approach) of Figure 3-1:

a) **Output Embedding**

Similar to the encoder, text is converted into numerical embeddings.

b) **Positional Encoding**

The same functionality as in the encoder.

c) **Decoder**

The decoder is also composed of a stack of six identical layers same as the encoder; the number of layers, depicted as Nx, can be increased or decreased. In addition to the two sublayers in each encoder layer, the decoder inserts a third sublayer, which performs multi-head attention over the output of the encoder stack.

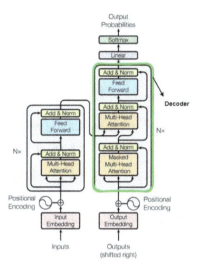

Figure 3-6. *The decoder highlighted in green*

d) **Masked Multi-head Self-Attention**

This layer is similar to the encoder's self-attention, but with a mask to prevent attending to future tokens or words. This masking plays a vital role in maintaining the order and coherence of the output.

e) **Encoder-Decoder Attention**

This layer allows the decoder to attend to the output of the encoder.

f) **Feedforward Neural Network**

Same as the encoder's feedforward network.

g) **Layer Normalization**

Applied after each sublayer to stabilize or improve the performance of the deep neural network.

h) **Residual Connections**

Used to ease training by allowing gradients to flow directly to earlier layers.

i) **Linear and Softmax**

The final layer converts the decoder output into predicted vocabulary or token probabilities.

This innovative approach revolutionized sequence modeling by introducing a transformer architecture that exclusively uses self-attention mechanisms, thereby eliminating the need for recurrent or convolutional neural networks, resulting in superior performance, parallelization, and accelerated training while effectively capturing long-range dependencies within data. There are different variants of transformers such as BERT, RoBERTa, LaMDA, GPT, Vit, T5, XLNet, and many more introduced for various applications.

3.2 Vanilla Transformer

Long sequence time series forecasting (LSTF) requires a high prediction capacity of the model, which can capture precise long-range dependency coupling between the output and the input efficiently; studies have shown the capabilities of transformers, but there are issues that prevent them from being directly applicable to LSTF, such as quadratic time complexity, high memory usage, and limitations of encoder-decoder architecture.

Transformer models have shown superior performance in capturing long-range dependency than RNN models; however, they still have some drawbacks such as

a) **The quadratic computation of self-attention**: The time complexity and memory usage per layer to be $O(L^2)$ when performing self-attention mechanisms, such as the canonical dot product.

b) **The memory bottleneck in stacking layers for long inputs**: The stack of J encoder/decoder layers makes total memory usage $O(J \cdot L^2)$, which limits the model scalability in receiving long sequence inputs.

c) **The speed reduction in predicting long outputs**: Dynamic decoding of a vanilla transformer makes the step-by-step inference as slow as regular sequence-based models, such as RNN, LSTM, or GRU.

3.2.1 Technical Overview of Vanilla Transformers

Vanilla transformers follow the implementation of an informer which is designed to handle long input sequences efficiently and capture complex patterns using ProbSparse attention, generative modeling, and self-attention, which helps mitigate traditional problems within transformers; the architecture has three distinct features:

a) Full-attention mechanism with O(L^2) time and memory complexity

b) Encoder-decoder with a multi-head attention mechanism as proposed by Vaswani et al. (2017) [1]

c) An MLP multi-step decoder that predicts long time series sequences in a single forward operation rather than step by step

3.2.2 What Is an Informer?

Informers are a specific type of neural network architecture; in other words, they improve over traditional transformers designed for long sequence time series forecasting and successfully enhance the prediction capacity in the LSTF problem, which proves the transformer-like model's potential value to capture individual long-range dependency between long sequence time series outputs and inputs. The key feature that is different from traditional transformers is ProbSparse.

Before understanding the ProbSparse self-attention mechanism, let's understand canonical self-attention, which is a variant of the self-attention mechanism explained earlier. In the traditional self-attention mechanism, we calculate attention scores between the elements (words) in the sequence. However, in canonical self-attention, we use convolutions which save computation time.

The **ProbSparse self-attention mechanism** is an advancement to the canonical self-attention. This technique aids the model to learn relationships between different parts of the input sequence with reduced memory and computational need.

The ProbSparse self-attention mechanism selects a subset of tokens for each query based on a probability distribution. This reduces the number of attention calculations, resulting in efficient computation for long sequences.

The self-attention distilling operation is a technique particularly used where a model selectively focuses on the most important attention weights within stacked layers (J-stacking layers). This process results in reduced memory footprint, which helps to receive long sequence input.

Pseudo code for ProbSparse self-attention

Require: Tensor $Q \in R^{m \times d}$, $K \in R^{n \times d}$, $V \in R^{n \times d}$

Q, K, V: Query, key, and value matrices, respectively

1: **print** set hyperparameter c, $u = c \ln m$ and $U = m \ln n$
2: randomly select U dot-product pairs from $\mathbf{K}$ as $\bar{\mathbf{K}}$
3: set the sample score $\bar{\mathbf{S}} = Q\bar{\mathbf{K}}^{\top}$
4: compute the measurement $M = \max(\bar{\mathbf{S}}) - \text{mean}(\bar{\mathbf{S}})$ by row
5: set Top-u queries under M as $\bar{\mathbf{Q}}$
6: set $\mathbf{S}_1 = \text{softmax}(\bar{\mathbf{Q}}\mathbf{K}^{\top}/\sqrt{d}) \cdot \mathbf{V}$
7: set $\mathbf{S}_0 = \text{mean}(\mathbf{V})$
8: set $\mathbf{S} = \{\mathbf{S}_1, \mathbf{S}_0\}$ by their original rows accordingly

Then the generative style decoder acquires long sequence output with only one forward step needed, simultaneously avoiding cumulative error spreading during the inference phase.

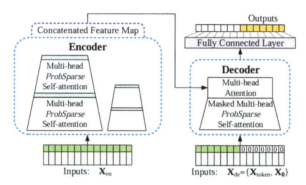

Figure 3-7. *Informer model overview [2]*

The encoder receives massive long sequence inputs X. We replace canonical self-attention with the proposed ProbSparse self-attention. The encoder block is the self-attention distilling operation to extract dominating attention, reducing the network size significantly. The layer stacking replicas increase robustness.

The decoder receives long sequence inputs, pads the target elements into zero, measures the weighted attention composition of the feature map, and instantly predicts output elements in a generative style.

The vanilla transformer model utilizes a three-component approach to define its embedding:

a) It uses encoded autoregressive features obtained from a convolution network.

b) It uses window-relative positional embeddings derived from harmonic functions which are a popular technique used in sequence modeling with transformers. Unlike absolute positional embeddings, which assign a unique embedding to each position in a sequence, window-relative embeddings focus on the relative positions of elements within a specific window. Harmonic functions can capture cyclical patterns and relationships between elements within a sequence. This technique is used for applications such as NLP and computer vision too.

c) Absolute positional embeddings are the vectors assigned to each position in the sequence of information about its location within the overall sequence. It is often beneficial to include calendar features directly into these embeddings, which helps create powerful embeddings that enhance the performance of your time series models.

3.2.3 Vanilla Transformer in Action

Having established a high-level theoretical foundation of a vanilla transformer, we shall now translate abstract concepts into practical code implementation.

Import required modules:

```
import numpy as np
import pandas as pd
import matplotlib.pyplot as plt
from neuralforecast import NeuralForecast
from neuralforecast.models import MLP,VanillaTransformer
from neuralforecast.losses.pytorch import MQLoss,
DistributionLoss,MAE
```

```
from neuralforecast.tsdataset import TimeSeriesDataset
from sklearn.metrics import mean_squared_error, mean_absolute_
percentage_error, r2_score
from neuralforecast.utils import AirPassengers,
AirPassengersPanel, AirPassengersStatic, augment_calendar_df
```

Let's load the dataset either from an offline copy or from the
neuralforecast.utils dataset, which contains 12 years of monthly air
passenger count. Separate the last 1 year of data for the test and use the
remaining 11 years of data to train the model.

```
from neuralforecast.utils import AirPassengersDF as Y_df
print(Y_df)
```

	unique_id	ds	y
0	1.0	1949-01-31	112.0
1	1.0	1949-02-28	118.0
2	1.0	1949-03-31	132.0
3	1.0	1949-04-30	129.0
4	1.0	1949-05-31	121.0

```
Y_train_df = Y_df[Y_df.ds<='1959-12-31']
Y_test_df = Y_df[Y_df.ds>'1959-12-31']
```

Let's construct and train the VanillaTransformer model by
understanding its key parameters.

h is the forecast horizon.

input size default is –1 which uses all the history, maximum sequence
length for truncated train backpropagation.

hidden_size is the unit of embeddings and encoders.

conv_hidden_size is the channels of the convolutional encoder.

n_heads is the number of multi-head attention.

scaler_type is the type of scaler for temporal input normalization.

learning_rate is the learning rate between (0, 1).

max_steps is the maximum number of training steps.

val_check_steps is the number of training steps between every validation loss check.

early_stop_patience_steps is the number of validation iterations before early stopping.

futr_exog_list, hist_exog_list, stat_exog_list are input parameters which can be used to add future, history, and static exogenous variables.

```
horizon = 12
model = VanillaTransformer(h=horizon,
                input_size=12,
                hidden_size=16,
                conv_hidden_size=32,
                n_head=2,
                loss=MAE(),
                #futr_exog_list=calendar_cols, example
                scaler_type='robust',
                learning_rate=1e-3,
                max_steps=500,
                val_check_steps=50,
                early_stop_patience_steps=2)

nf = NeuralForecast(
    models=[model],
    freq='M'
)
```

Let's train the model using training data, and val_size is the validation size for temporal cross-validation:

```
nf.fit(df=Y_train_df, val_size=12)
```

Predict the next defined horizon, which is 12 months:

```
forecasts = nf.predict()
forecasts.head()
```

	ds	VanillaTransformer
unique_id		
1.0	1960-01-31	399.015778
1.0	1960-02-29	419.278839
1.0	1960-03-31	446.802612
1.0	1960-04-30	446.806213
1.0	1960-05-31	449.698669

Let's measure the model's accuracy:

```
calculate_error_metrics(Y_test_df[['y']],forecasts
['VanillaTransformer'])
```

```
MSE : 688.7931518554688
RMSE : 26.244869232177734
MAPE : 0.04862939938902855
r2 : 0.8756572681442947
adjusted_r2 : 0.8632229949587241
```

Visualize actual vs. predicted:

```
Y_train_df.set_index('ds',inplace =True)
forecasts.set_index('ds',inplace =True)
Y_test_df.set_index('ds',inplace =True)
plt.figure(figsize=(20, 3))
y_past = Y_train_df["y"]
y_pred = forecasts['VanillaTransformer']
y_test = Y_test_df["y"]
plt.plot(y_past, label="Past time series values")
plt.plot(forecasts, label="Forecast")
plt.plot(y_test, label="Actual time series values")
plt.title('AirPassengers Forecast', fontsize=10)
plt.ylabel('Monthly Passengers', fontsize=10)
plt.xlabel('Timestamp [t]', fontsize=10)
plt.legend();
```

Figure 3-8 helps us to appreciate that the air passenger count predicted by our model is close to reality.

Figure 3-8. *Actual vs. predicted plot*

3.3 Inverted Transformers

The recent popularity of linear forecasting models questions the ongoing interest in architectural modifications of transformer-based forecasters. Transformer-based forecasting typically embeds multiple variates of the same timestamp into almost identical channels and applies attention on these temporal tokens to capture temporal dependencies.

Transformers struggle in forecasting series with larger look-back windows due to performance degradation and computation explosion; however, the embedding for each temporal token fuses multiple variates that represent potential delayed events and distinct physical measurements, which may fail in learning variate-centric representations and result in meaningless attention maps. The main idea here is to reuse the transformer without any modification to the basic components.

3.3.1 Technical Overview of iTransformers

iTransformers take the transformer architecture but apply the attention and feedforward network on the inverted dimensions. This means that the time points of each individual series are embedded into variate tokens which can be used by the attention mechanisms to capture multivariate correlation, and the feedforward network learns nonlinear relationships.

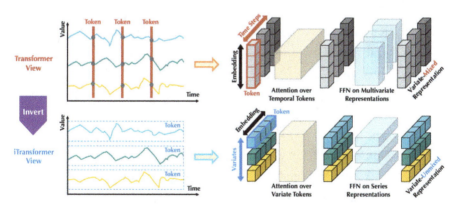

Figure 3-9. *Comparison between the vanilla transformer (top) and the iTransformer (bottom) [3]*

The transformer embeds the temporal token, which contains the multivariate representation of each time step.

103

The iTransformer embeds each series independently to the token, such that the attention module depicts the multivariate correlations and the feedforward network encodes series representations.

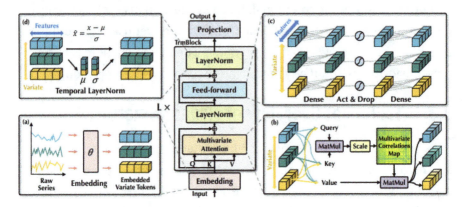

Figure 3-10. *Structure from iTransformers [3]*

a) **Embedding**

Raw series of different variates are independently embedded as variate tokens and passed to the next step.

b) **Self-attention**

The inverted model regards the whole series of one variate as an independent process. Concretely, with comprehensively extracted representations of each time series $H = \{h_0,...,h_N\} \in R^{N \times D}$, the self-attention module adopts linear projections to get queries, keys, and values Q, K, V $\in R^{N \times d_k}$, where d_k is the projected dimension. Self-attention is applied to embedded variate tokens with enhanced interpretability revealing multivariate correlations.

c) **Feedforward network**

The feedforward network is leveraged on the series representation of each variate token. By the universal approximation theorem, they can extract complicated representations to describe a time series such as amplitude, periodicity, and even frequency spectrums. With the stacking of inverted blocks, they are devoted to encoding the observed time series and decoding the representations for future series using dense nonlinear connections.

d) **Layer normalization**

The normalization is applied to the series representation of individual variate as the equation mentioned below; since all series as (variate) tokens are normalized to a Gaussian distribution, the discrepancies caused by inconsistent measurements can be diminished and hence adopted to reduce the discrepancies among variates.

$$\mathrm{LayerNorm}(\mathbf{H}) = \left\{ \left. \frac{\mathbf{h}_n - \mathrm{Mean}(\mathbf{h}_n)}{\sqrt{\mathrm{Var}(\mathbf{h}_n)}} \right| n = 1, \ldots, N \right\}$$

3.3.2 iTransformers in Action

Having established a high-level theoretical foundation of inverted transformer, we shall now translate abstract concepts into practical code implementation.

Import required modules:

```
import numpy as np
import pandas as pd
import matplotlib.pyplot as plt
```

```
from neuralforecast import NeuralForecast
from neuralforecast.models import iTransformer
from neuralforecast.losses.pytorch import MQLoss,
DistributionLoss,MSE
from neuralforecast.tsdataset import TimeSeriesDataset
from sklearn.metrics import mean_squared_error, mean_absolute_
percentage_error, r2_score
```

Let's load the dataset either from an offline copy or from the **neuralforecast.utils** dataset, which contains 12 years of monthly air passenger count. Separate the last 1 year of data for the test and use the remaining 11 years of data to train the model.

```
from neuralforecast.utils import AirPassengersDF as Y_df
Y_train_df = Y_df[Y_df.ds<='1959-12-31']
Y_test_df = Y_df[Y_df.ds>'1959-12-31']
dataset, *_ = TimeSeriesDataset.from_df(Y_train_df)
```

Let's construct and train the iTransformer model by understanding its key parameters.

h is the horizon.

input_size is the autoregressive input size, y=[1,2,3,4] input_size=2 -> y_[t-2:t]=[1,2].

n_series is the number of time series.

hidden_size is the dimension of the model.

n_heads is the number of heads.

e_layers is the number of encoder layers.

d_layers is the number of decoder layers.

d_ff is the dimension of the fully connected layer.

factor is the attention factor.

dropout is the dropout rate.

use_norm is whether to normalize or not.

loss is the instantiated train loss class from the losses collection.

valid_loss is the instantiated valid loss class from the losses collection.

batch_size is the number of different series in each batch.

futr_exog_list, hist_exog_list, stat_exog_list are input parameters which can be used to add future, history, and static exogenous variables.

```
horizon =12
model =  iTransformer(h=horizon,
                      input_size=24,
                      n_series=2,
                      hidden_size=128,
                      n_heads=2,
                      e_layers=2,
                      d_layers=1,
                      d_ff=4,
                      factor=1,
                      dropout=0.1,
                      use_norm=True,
                      loss=MSE(),
                      valid_loss=MAE(),
                      batch_size=32)
model.fit(dataset=dataset,val_size=12)
```

Predict the next defined horizon, which is 12 months:

```
y_hat = model.predict(dataset=dataset)
Y_test_df['iTransformers'] = y_hat
Y_test_df.head()
```

	unique_id	ds	y	iTransformers
132	1.0	1960-01-31	417.0	425.598022
133	1.0	1960-02-29	391.0	390.221741
134	1.0	1960-03-31	419.0	435.384155
135	1.0	1960-04-30	461.0	423.194702
136	1.0	1960-05-31	472.0	460.768036

Let's measure the model's accuracy:

```
calculate_error_metrics(Y_test_df[['y']],Y_test_
df['iTransformers'])
```

```
MSE : 335.6207580566406
RMSE : 18.319955825805664
MAPE : 0.030659830197691917
r2 : 0.9394128604587846
adjusted_r2 : 0.9333541465046631
```

Visualize actual vs. predicted:

```
Y_train_df.set_index('ds',inplace =True)
Y_test_df.set_index('ds',inplace =True)
plt.figure(figsize=(20, 3))
y_past = Y_train_df["y"]
y_pred = Y_test_df['iTransformers']
y_test = Y_test_df["y"]
plt.plot(y_past, label="Past time series values")
plt.plot(y_pred, label="Forecast")
plt.plot(y_test, label="Actual time series values")
```

```
plt.title('AirPassengers Forecast', fontsize=10)
plt.ylabel('Monthly Passengers', fontsize=10)
plt.xlabel('Timestamp [t]', fontsize=10)
plt.legend();
```

Figure 3-11 helps us to appreciate that the air passenger count predicted by our model is close to reality.

Figure 3-11. *Actual vs. predicted*

3.4 DLinear

Forecasting a larger horizon is only feasible for those time series with a clear trend and cyclicity, as linear models can readily extract such information. Simple models such as long-term time series forecasting LTSF-Linear regress historical time series with a one-layer linear model to forecast. Results show that LTSF-Linear outperforms existing complex transformer models in all cases by a large margin.

Moreover, most of the transformers fail to extract temporal relations which are connections between events related to each other from long sequences. When such sequences occur, the forecasting errors are not reduced (sometimes even increased) with the increase in look-back window sizes.

LTSF-Linear is a set of linear models. Vanilla Linear is a one-layer linear model to handle time series across different domains (e.g., weather forecast, retail, and healthcare); we further understand two variants with two preprocessing methods, named DLinear and NLinear.

The architecture has the following distinctive features: uses Autoformer's trend and seasonality decomposition and simple linear layers for trend and seasonality components.

The LTSF-Linear directly regresses historical time series for future prediction via a weighted sum operation (Figure 3-12).

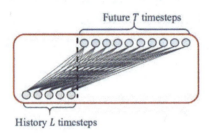

Figure 3-12. *Illustration of one basic linear layer [4]*

The mathematical expression is $Xw_i = WX_i$, where $W \in R^{T \times L}$ is a linear layer along the temporal axis. $\hat{X}_i$ and X_i are the prediction and input for each i_{th} variate. LTSF-Linear shares weights across different variates and does not model any spatial correlations.

3.4.1 What Is Autoformer ?

The Autoformer model is based on the decomposition of time series into seasonality and trend cycle. To achieve this, a decomposition layer is added, which enhances the model's ability. Autoformer uses an innovative auto-correlation mechanism that enables the model to utilize period-based dependencies in the attention. This improves accuracy in finding reliable dependencies on intricate temporal patterns of long-horizon forecasting.

3.4.2 Technical Overview of DLinear

DLinear is a combination of a decomposition scheme used in Autoformer and FEDformer with linear layers. It decomposes raw data input into a trend component by a moving average kernel and a remainder or seasonal component. Then, two one-layer linear layers are applied to each component, and we sum up the two features to get the final prediction. By explicitly handling trends, DLinear enhances the performance of a vanilla linear when there is a clear trend in the data.

This architecture has the following unique features compared to traditional architecture:

- Built-in progressive decomposition in trend and seasonal components based on a moving average filter where decomposed components are updated and refined iteratively during the forecasting process. This is a dynamic process compared to traditional decomposition where the decomposition of components is fixed throughout the forecasting process.

- The autocorrelation mechanism discovers the period-based dependencies by calculating the autocorrelation and aggregating similar subseries based on the periodicity.

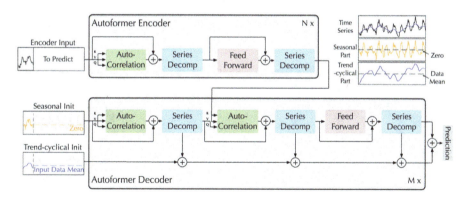

Figure 3-13. *Autoformer architecture [5]*

a) **Encoder**

The encoder focuses on seasonal part of the modeling which helps decoder to use use the past seasonal information generated by encoder to refine the prediction.

b) **Decoder**

The decoder performs well in two tasks, which are an accumulation of the structure of trend-cyclical components and the stacked autocorrelation mechanism for season components shown in Figure 3-13.

Each decoder is comprised of inner autocorrelation and encoder-decoder autocorrelation which helps in refining prediction by utilizing the past seasonal information. The model extracts the trend components from the intermediate hidden variables during the decoder, allowing Autoformer to progressively refine the trend prediction and eliminate interference information for period-based dependency discovery in autocorrelation.

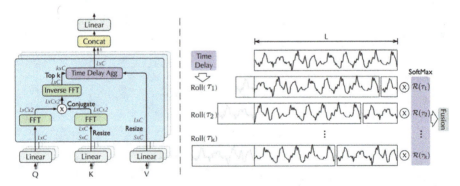

Figure 3-14. *Autocorrelation (left) and time delay aggregation (right) [5]*

c) **Autocorrelation**

The autocorrelation mechanism with series-wise connections to expand the information utilization. Autocorrelation discovers the period-based dependencies by calculating the series autocorrelation and aggregates similar subseries by time delay aggregation.

The Fast Fourier Transform is used to calculate the autocorrelation R(T), which reflects the time delay similarities. Then the similar subprocesses are rolled to the same index based on selected delay T and aggregated by R(T). The final prediction is the sum of the two refined decomposed components.

d) **Time delay aggregation**

The period-based dependencies connect the subseries among estimated periods, as depicted in the time delay aggregation block in Figure 3-14 (right). Time delay aggregation can roll the series based on the selected time delay. This operation can align similar subseries that are at the same phase position of estimated periods. This is different from the point-wise dot-product aggregation in the self-attention family. Finally, it aggregates the subseries by softmax normalized confidences.

113

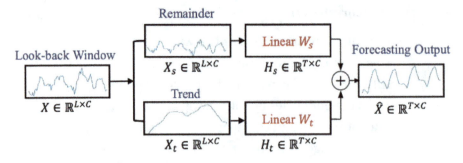

Figure 3-15. *Structure of DLinear [4]*

$\hat{X} = H_s + H_t$, where $H_s = W_s X_s \in R^{T \times C}$ are the decomposed trend and remainder features.

$W_s \in R^{T \times L}$ and $W_t \in R^{T \times L}$ are the two linear layers.

DLinear is capable of capturing both short-range and long-range temporal relations, and as each branch has only one linear layer, it costs much lower memory and fewer parameters and has a faster inference speed than existing transformers.

3.4.3 DLinear in Action

Having established a high-level theoretical foundation of DLinear, we shall now translate abstract concepts into practical code implementation.

Import required libraries:

```
import numpy as np
import pandas as pd
import matplotlib.pyplot as plt
from neuralforecast import NeuralForecast
from neuralforecast.models import MLP,DLinear
from neuralforecast.losses.pytorch import MQLoss,
DistributionLoss,MAE
```

```
from neuralforecast.tsdataset import TimeSeriesDataset
from sklearn.metrics import mean_squared_error, mean_absolute_
percentage_error, r2_score
from neuralforecast.utils import AirPassengers,
AirPassengersPanel, AirPassengersStatic, augment_calendar_df
```

Let's load the dataset either from an offline copy or from the **neuralforecast.utils** dataset, which contains 12 years of monthly air passenger count. Separate the last 1 year of data for the test and use the remaining 11 years of data to train the model.

```
from neuralforecast.utils import AirPassengersDF
Y_df = AirPassengersDF
Y_df = Y_df.reset_index(drop=True)
Y_train_df = Y_df[Y_df.ds<='1959-12-31']
Y_test_df = Y_df[Y_df.ds>'1959-12-31']
```

Let's train the DLinear model and define its hyperparameters.

h is the horizon.

input_size is the autoregressive input size, y=[1,2,3,4] input_size=2 -> y_[t-2:t]=[1,2].

loss is the instantiated train loss class from the losses collection.

Scaler_type is the type of scaler for temporal input normalization.

Learning_rate is the learning rate between (0, 1).

max_steps is the maximum number of training steps.

val_check_steps is the number of training steps between every validation loss check.

early_stop_patience_steps is the number of validation iterations before early stopping.

futr_exog_list, hist_exog_list, stat_exog_list are input parameters which can be used to add future, history, and static exogenous variables.

```
horizon =12
model = DLinear(h=horizon,
                input_size=12,
                loss=MAE(),
                scaler_type='robust',
                learning_rate=1e-3,
                max_steps=500,
                val_check_steps=50,
                early_stop_patience_steps=2)

nf = NeuralForecast(
    models=[model],
    freq='M'
)
nf.fit(df=Y_train_df, val_size=12)
```

Predict the next defined horizon:

```
forecasts = nf.predict()
forecasts.head()
```

	ds	DLinear
unique_id		
1.0	1960-01-31	429.155396
1.0	1960-02-29	452.166656
1.0	1960-03-31	411.675354
1.0	1960-04-30	406.123108
1.0	1960-05-31	499.705017

Measure the model accuracy:

```
calculate_error_metrics(Y_test_df[['y']],forecasts['DLinear'])
```

```
MSE : 1411.2696533203125
RMSE : 37.56686782836914
MAPE : 0.07227641344070435
r2 : 0.7452339051869932
adjusted_r2 : 0.7197572957056925
```

Visualize the predictions:

```
Y_train_df.set_index('ds',inplace =True)
forecasts.set_index('ds',inplace =True)
Y_test_df.set_index('ds',inplace =True)
plt.figure(figsize=(20, 3))
y_past = Y_train_df["y"]
y_pred = forecasts['DLinear']
y_test = Y_test_df["y"]
plt.plot(y_past, label="Past time series values")
plt.plot(y_pred, label="Forecast")
plt.plot(y_test, label="Actual time series values")
plt.title('AirPassengers Forecast', fontsize=10)
plt.ylabel('Monthly Passengers', fontsize=10)
plt.xlabel('Timestamp [t]', fontsize=10)
plt.legend();
```

Figure 3-16 helps us to appreciate that the air passenger count predicted by our model is close to reality.

Figure 3-16. *Actual vs. predicted*

3.5 NLinear

NLinear is part of the LTSF-Linear family of models specifically designed to boost the performance of Linear when there is a distribution shift in the dataset. NLinear first subtracts the input by the last value of the sequence, then the input goes through a linear layer, and the subtracted part is added back before making the final prediction. The subtraction and addition in NLinear are a simple normalization for the input sequence.

NLinear can consistently outperform all transformer-based methods by a large margin most of the time. Simple normalization via the last value from the look-back window can greatly relieve the distribution shift problem.

A distribution shift occurs when the statistical properties of the training data differ significantly from the test data. This is when the model is trained on one set of data but applied to data of different characteristics. The various types of distribution shifts are covariate shift, label shift, and concept drift. If these kinds of shifts are not handled properly, it can result in performance degradation and unreliable predictions.

3.5.1 NLinear in Action

Having established a high-level foundation of NLinear, we shall now translate abstract concepts into practical code implementation.

Import required modules:

```
import numpy as np
import pandas as pd
import matplotlib.pyplot as plt
from neuralforecast import NeuralForecast
from neuralforecast.models import MLP,NLinear
from neuralforecast.losses.pytorch import MQLoss,
DistributionLoss,MAE
from neuralforecast.tsdataset import TimeSeriesDataset
from sklearn.metrics import mean_squared_error, mean_absolute_
percentage_error, r2_score
from neuralforecast.utils import AirPassengers,
AirPassengersPanel, AirPassengersStatic, augment_calendar_df
```

Let's load the dataset either from an offline copy or from the **neuralforecast.utils** dataset, which contains 12 years of monthly air passenger count. Separate the last 1 year of data for the test and use the remaining 11 years of data to train the model.

```
from neuralforecast.utils import AirPassengersDF
Y_df = AirPassengersDF
Y_df = Y_df.reset_index(drop=True)
Y_df.head()

Y_train_df = Y_df[Y_df.ds<='1959-12-31']
Y_test_df = Y_df[Y_df.ds>'1959-12-31']
```

Let's train the NLinear model and define its hyperparameters.
h is the horizon.
input_size is the autoregressive input size, y=[1,2,3,4] input_size=2 -> y_[t-2:t]=[1,2].

119

loss is the instantiated train loss class from the losses collection.

Scaler_type is the type of scaler for temporal input normalization.

Learning_rate is the learning rate between (0, 1).

max_steps is the maximum number of training steps.

val_check_steps is the number of training steps between every validation loss check.

early_stop_patience_steps is the number of validation iterations before early stopping.

futr_exog_list, hist_exog_list, stat_exog_list are input parameters which can be used to add future, history, and static exogenous variables.

```
horizon =12
model = NLinear(h=horizon,
                input_size=12,
                loss=MAE(),
                scaler_type='robust',
                learning_rate=1e-3,
                max_steps=500,
                val_check_steps=50,
                early_stop_patience_steps=2)

nf = NeuralForecast(
    models=[model],
    freq='M'
)

nf.fit(df=Y_train_df, val_size=12)
```

Predict the next defined horizon:

```
forecasts = nf.predict()
forecasts.head()
```

unique_id	ds	NLinear
1.0	1960-01-31	419.600891
1.0	1960-02-29	446.448914
1.0	1960-03-31	434.517181
1.0	1960-04-30	434.610718
1.0	1960-05-31	483.879761

Measure the model accuracy:

```
calculate_error_metrics(Y_test_df[['y']],forecasts['NLinear'])
```

```
MSE : 1214.2550048828125
RMSE : 34.846160888671875
MAPE : 0.06048479676246643
r2 : 0.7807995266526639
adjusted_r2 : 0.7588794793179303
```

Visualize the predictions:

```
Y_train_df.set_index('ds',inplace =True)
forecasts.set_index('ds',inplace =True)
Y_test_df.set_index('ds',inplace =True)
plt.figure(figsize=(20, 3))
y_past = Y_train_df["y"]
y_pred = forecasts['NLinear']
y_test = Y_test_df["y"]
plt.plot(y_past, label="Past time series values")
plt.plot(y_pred, label="Forecast")
```

```
plt.plot(y_test, label="Actual time series values")
plt.title('AirPassengers Forecast', fontsize=10)
plt.ylabel('Monthly Passengers', fontsize=10)
plt.xlabel('Timestamp [t]', fontsize=10)
plt.legend();
```

Figure 3-17 helps us to appreciate that the air passenger count predicted by our model is close to reality.

Figure 3-17. *Actual vs. predicted*

3.6 Patch Time Series Transformer

PatchTST supports multivariate time series forecasting and self-supervised representation learning. It is based on the segmentation of time series into subseries-level patches, which serve as input tokens to the transformer.

Channel independence is a property of PatchTST where each channel contains a single univariate time series. Channel independence helps share the same embedding and transformer weights across all the series. This helps the PatchTST model to apply attention weights separately to each channel, which helps in better capturing the unique features and patterns in each channel.

Patching is the segmentation of time series into windows, which helps to enhance the locality and capture comprehensive semantic information that is not available at the point level. This is achieved by aggregating time steps into subseries-level patches and channel independence.

3.6.1 Technical Overview of PatchTST

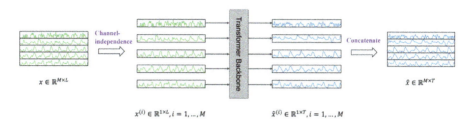

Figure 3-18. *PatchTST model overview [6]*

Multivariate time series data is divided into different channels. They share the same transformer backbone, but the forward processes are independent.

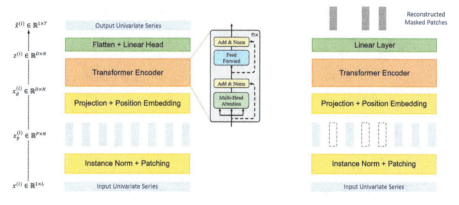

Figure 3-19. *(a) Transformer backbone (supervised), (b) transformer backbone (self-supervised) [6]*

- Each channel univariate series is passed through the instance normalization operator and segmented into patches. These patches are used as transformer input tokens.

- Masked self-supervised representation learning with PatchTST where patches are randomly selected and set to zero. The model will reconstruct the masked patches.

a) **Patching**

Point-wise attention which is used in traditional transformers tries to retrieve information from a single step. This is not ideal in a time series, as we will need to extract the relationship between past time steps and future time steps to make predictions.

Univariate inputs of time series $x^{(i)}$ are first divided into patches which can be either overlapped or non-overlapped. Denote the patch length as P and the stride – the non-overlapping region between two consecutive patches – as S, then the patching process will generate the sequence of patches $x_p^{(i)}$ $\in R^{P \times N}$ where N is the number of patches, $N = [($ $L-P)/ S] +2$. Here, we pad S repeated numbers of the last value $x_L^{(i)} \in R$ to the end of the original sequence before patching.

The number of input tokens can be reduced with the use of patches from L to approximately L/S. This implies the memory usage and computational complexity of the attention map are quadratically decreased by a factor of S.

b) **Transformer encoder**

A vanilla transformer encoder is used that maps the observed signals to the latent representations. The patches are mapped to the transformer latent space of dimension D via a trainable linear projection, and a learnable additive position encoding is applied to monitor the temporal order of patches.

c) **Loss function**

MSE loss to measure the discrepancy between the prediction and the ground truth. The loss in each channel is gathered and averaged over M time series to get the overall objective loss:

$$\mathcal{L} = \mathbb{E}_x \frac{1}{M} \sum_{i=1}^{M} \left\| \hat{x}^{(i)}_{L+1:L+T} - x^{(i)}_{L+1:L+T} \right\|_2^2.$$

d) **Instance normalization**

This helps mitigate the distribution shift effect between the training and testing data. It simply normalizes each time series instance x$^{(i)}$ with zero mean and unit standard deviation. In this type of normalization, for each x$^{(i)}$ before patching, the mean and deviation are added back to the output prediction.

Representation Learning

PatchTST can be utilized for self-supervised use cases to capture the abstract representation of the data. The same encoder is used as the supervised settings, the prediction head is removed, and a D×P linear layer is attached. Instead of a supervised model where patches can be overlapped, here each input sequence is split into regular non-overlapping patches.

It is for convenience to ensure observed patches do not contain information about the masked patches. This is achieved by selecting a subset of the patch at random and masking the patches according to zero values. The model is trained with MSE loss to reconstruct the masked patches.

3.6.2 PatchTST in Action

Having established a high-level theoretical foundation of the PatchTST transformer, we shall now translate abstract concepts into practical code implementation.

```python
import numpy as np
import pandas as pd
import matplotlib.pyplot as plt
from neuralforecast import NeuralForecast
from neuralforecast.models import PatchTST
from neuralforecast.losses.pytorch import MQLoss,
DistributionLoss,MAE
from neuralforecast.tsdataset import TimeSeriesDataset
from sklearn.metrics import mean_squared_error, mean_absolute_
percentage_error, r2_score
from neuralforecast.utils import AirPassengers,
AirPassengersPanel, AirPassengersStatic, augment_calendar_df
```

Let's load the dataset either from an offline copy or from the **neuralforecast.utils** dataset, which contains 12 years of monthly air passenger count. Separate the last 1 year of data for the test and use the remaining 11 years of data to train the model.

```python
from neuralforecast.utils import AirPassengersDF as Y_df
Y_train_df = Y_df[Y_df.ds<='1959-12-31'] # 132 train
Y_test_df = Y_df[Y_df.ds>'1959-12-31']   # 12 test

dataset, *_ = TimeSeriesDataset.from_df(Y_train_df)
```

Let's initialize and train the PatchTST model:

h is the horizon.

input_size is the autoregressive input size, y=[1,2,3,4] input_size=2 -> y_[t-2:t]=[1,2].

Patch_length is the length of patch. Note: patch_len = min(patch_len, input_size + stride).

Stride is the stride of patch.

revin is the RevIn.

hidden_size is the dimension of the model.

n_heads is the number of heads.

scaler_type is the type of scaler for temporal input normalization; see temporal scalers.

loss is the instantiated train loss class from the losses collection.

learning_rate is the learning rate between (0, 1).

max_steps is the maximum number of training steps.

val_check_steps is the number of training steps between every validation loss check.

early_stop_patience_steps is the number of validation iterations before early stopping.

futr_exog_list, hist_exog_list, stat_exog_list are input parameters which can be used to add future, history, and static exogenous variables.

```python
horizon =12
model = PatchTST(h=horizon,
                 input_size=104,
                 patch_len=12,
                 stride=24,
                 revin=False,
                 hidden_size=16,
                 n_heads=4,
                 scaler_type='robust',
                 loss=DistributionLoss(distribution='StudentT',
                 level=[80, 90]),
                 learning_rate=1e-3,
                 max_steps=500,
                 val_check_steps=50,
                 early_stop_patience_steps=2)

nf = NeuralForecast(
    models=[model],
    freq='M'
)
nf.fit(df=Y_train_df, static_df=AirPassengersStatic, val_size=12)
```

Predict the next defined horizon:

```
forecasts = nf.predict()
forecasts.head()
```

The predicted column of interest is PatchTST.

	ds	PatchTST	PatchTST-median	PatchTST-lo-90	PatchTST-lo-80	PatchTST-hi-80	PatchTST-hi-90
unique_id							
1.0	1960-01-31	398.039093	398.154114	371.832336	378.644867	416.888031	423.438202
1.0	1960-02-29	372.010956	371.771698	339.672333	346.152863	396.310883	405.729370
1.0	1960-03-31	408.209717	408.825989	364.338135	376.066437	439.823029	451.888763
1.0	1960-04-30	412.931366	413.688782	382.853210	391.648163	434.949524	441.165405
1.0	1960-05-31	416.384338	418.396515	367.197998	383.222198	448.187531	458.514557

Measure the model accuracy:

```
calculate_error_metrics(Y_test_df[['y']],forecasts['PatchTST'])
```

```
MSE : 1601.8538818359375
RMSE : 40.02316665649414
MAPE : 0.07522071897983551
r2 : 0.7108291425217838
adjusted_r2 : 0.6819120567739622
```

Visualize the predictions:

```
Y_train_df.set_index('ds',inplace =True)
forecasts.set_index('ds',inplace =True)
Y_test_df.set_index('ds',inplace =True)
plt.figure(figsize=(20, 3))
y_past = Y_train_df["y"]
y_pred = forecasts['PatchTST']
y_test = Y_test_df["y"]
plt.plot(y_past, label="Past time series values")
plt.plot(y_pred, label="Forecast")
plt.plot(y_test, label="Actual time series values")
```

```
plt.title('AirPassengers Forecast', fontsize=10)
plt.ylabel('Monthly Passengers', fontsize=10)
plt.xlabel('Timestamp [t]', fontsize=10)
plt.legend();
```

Figure 3-20 helps us to appreciate that the air passenger count predicted by our model is close to reality.

Figure 3-20. *Actual vs. predicted*

3.7 Summary

In this chapter, we explored the transformer architecture, its core components, and the modifications to leverage this powerful architecture for time series forecasting tasks. We also discussed other variants such as vanilla transformers, inverted transformers, DLinear, NLinear, and PatchTST. We gained insights into the strengths and weaknesses of different approaches, equipping readers to make informed decisions when selecting models for specific time series forecasting tasks.

Choosing the best model depends on multiple parameters like distributions of features, properties (volume, missing values, number of features, etc.) of the dataset, and parameters like cost effectiveness, memory usage, and computation power. However, we now understand the strengths and weaknesses of the models discussed.

Vanilla transformers are highly scalable and versatile due to their ability to be used in different domains like NLP, computer vision, and time series. On the downside, they come with a high computational cost and are data hungry. They are also not the best choice for use cases with continuous time series data.

129

Inverted transformer architecture is best suited for sequential data while limited to time series use cases and requires expertise in hyperparameter tuning.

DLinear architecture needs lesser features to train and has lower memory and computational needs. These are the go-to architectures in resource-constrained use cases. However, this model is not best when there are nonlinear relationships in the data. The NLinear model is useful in cases where the data is stationary or nonstationary; however, it does not capture complex relations within the data.

Finally, the PatchTST captures the dependencies in the data well while still having downsides with respect to preprocessing needs and careful patch size selection.

3.8 References

[1].   Attention is all you need. https://doi.org/10.48550/arXiv.1706.03762

[2].   iTransformer: Inverted Transformers Are Effective for Time Series Forecasting. https://doi.org/10.48550/arXiv.2310.06625

[3].   Informer: Beyond Efficient Transformer for Long Sequence Time-Series Forecasting. https://doi.org/10.48550/arXiv.2012.07436

[4].   Are Transformers Effective for Time Series Forecasting? https://doi.org/10.48550/arXiv.2205.13504

[5].   Autoformer: Decomposition Transformers with Auto-Correlation for Long-Term Series Forecasting. https://doi.org/10.48550/arXiv.2106.13008

[6].   A Time Series is Worth 64 Words: Long-term Forecasting with Transformers. https://doi.org/10.48550/arXiv.2211.14730

CHAPTER 4

Time-LLM: Reprogramming Large Language Model

Chapter Goal: Understand how the Time-LLM repurposes a foundation model that is designed for NLP tasks and uses it for time series forecasting.

In the previous chapter, we covered transformers. In the upcoming chapters, we will understand how to use large language models built with the help of transformers. We will discuss some of the recent foundation models used in time series forecasting, starting with the first one: TimeGPT. Some of the recent advances in time series foundation models make use of techniques like fine-tuning or pre-training to capture generalized knowledge for time series forecasting.

In this chapter, we will cover Time-LLM. In Chapter 2, we discussed the WaveNet model. While it was designed primarily for audio applications, we saw how to use it for time series forecasting by changing parameters like dilation rate, receptive field, and loss function. In this chapter, let us discuss how a foundation model primarily trained and used for NLP applications can be used for time series forecasting. Time-LLM essentially provides a framework to tackle this challenge, without changing the model itself. In this framework, the input time series is transformed to a natural

© Banglore Vijay Kumar Vishwas and Sri Ram Macharla 2025
B. V. Vishwas and Sri Ram Macharla, *Time Series Forecasting Using Generative AI*,
https://doi.org/10.1007/979-8-8688-1276-7_4

language before feeding it to the foundation model. The output of the foundation model is then decoded to a time series forecast.

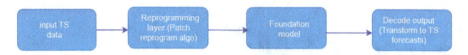

Figure 4-1. *High-level overview of Time-LLM*

Before proceeding to understand this framework, we need to clearly understand the difference between reprogramming and fine-tuning. While both methods help to adopt foundation models to perform desired tasks, they differ in process and purpose.

4 Fine-Tuning vs. Reprogramming

Fine-tuning involves extending the training process on the foundation model rather than from scratch. Not everyone has the resources and time to build a custom model. The process of fine-tuning involves training the model with our custom data. This helps to serve a specific task or to use in a specific domain. The process of training generally may be achieved by freezing some layers or by training with a different learning rate. The weights of the model are adjusted, resulting in a change in the nature of the foundation model itself. One of the use cases where we used fine-tuning is taking a foundation model like GPT-2/BERT and fine-tuning to perform sentiment analysis. This was part of onboarding a new client based on media articles, reviews, and a few other written sources.

Reprogramming involves using a foundation model built for use in a specific domain for an entirely different task or domain. This is generally achieved by using a transformation layer or a mapping layer to convert input from a different domain to the domain that the model understands. For example, input time series data to NLP data. It focuses on altering input rather than altering the model itself. This is similar to using wrappers in software engineering. This is faster and less resource intensive since the effort is in translating the input than the core model itself. This translation layer can be reused for similar input transformation (input domain -> output domain mappings), which is not necessarily true for fine-tuning cases.

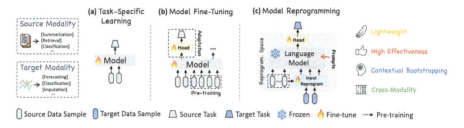

Figure 4-2. *Visual representation of fine-tuning (b) and reprogramming (c) [1]*

From the above understanding, it is clear that fine-tuning changes the model parameters to cater to specific tasks, whereas reprogramming helps us to use the model for an entirely different task, by presenting the input in a format that the model was primarily intended to accept, without changing (or minimal changes) to the model itself.

4.1 Technical Overview of Time-LLM

Let us understand the enhancements that can be done to a foundation model primarily trained on text data, like GPT-2, to be used for time series forecasting. The foundation model works by converting input text to multidimensional vectors that are used to capture the semantic properties of the input text. Figure 4-3 represents the framework that we are about to discuss.

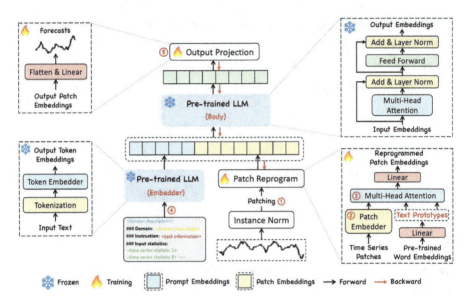

Figure 4-3. *Model framework of Time-LLM [1]*

Figure 4-3 shows the model framework of Time-LLM. The input time series is converted to tokens, and embedding is performed by (1) patching along with a (2) customized embedding layer. (3) These patch embeddings are then reprogrammed with condensed text prototypes to align two modalities. To augment the LLM's reasoning ability, (4) additional prompt prefixes are added to the input to direct the transformation of input patches. (5) The output patches from the LLM are passed through a projection layer to generate the forecasts.

4.1.1 Working of Time-LLM

Each input time series is individually normalized, which results in a time series with unit SD (standard deviation) and zero mean. This is achieved using the RevIN (reversible input normalization) algorithm. RevIN is a type of normalization where the transformations can be rolled back, thus helping to recover original data after processing. Then the input time series is divided into many overlapped or non-overlapped patches, each of length L_p. Patching of input time series helps in preserving the local semantic information. LSI is nothing but understanding the words (or n-grams) within their immediate context. This is achieved by considering a group of time steps rather than an individual time step. This aggregation of time steps results in greatly reducing the number of tokens passed to the reprogramming layer. This is due to the generation of a compact sequence of input tokens for reprogramming that greatly reduces computational complexity. Patch embeddings are generated as a result of this step.

The patch embeddings generated from the first (previous) step are used in patch reprogramming. This step innovatively transforms the patch embeddings to text prototypes. This helps the foundation model trained with natural language to understand time series data. This is achieved by taking help of techniques used in domain adaptation like noise learning. Here, instead of retraining, a small noise (perturbation) is learned. This noise when applied to the input patch embeddings generates output that can be understood by the foundation model.

Those familiar with domain adaptation may be already thinking; it is possible to achieve tasks within the same domain like models trained on images in day light to identify images taken in poor lighting conditions by introducing noise in the form of brightness and contrast. However, text and numbers are entirely different domains. To handle this unique scenario, the patch reprogramming layer leverages pre-trained word embeddings already present in the foundation model. Another challenge arises in using since there is no information regarding the relevance of source tokens. Leveraging all possible word embeddings results in a large co-domain. This is overcome by selecting a small subset of relevant word embeddings by using a linear classifier (like a logistic regression or linear SVC) to classify relevant embeddings for our task. To understand better, we can see that in Figure 4-4 text prototypes learn connecting language cues, for example, "short up" (shown in red lines) "steady down" (shown in blue lines). These are combined to represent the local patch information "short up then down steadily" for characterizing patch 5. Similarly, "early down," "steady long" for characterizing patch 1.

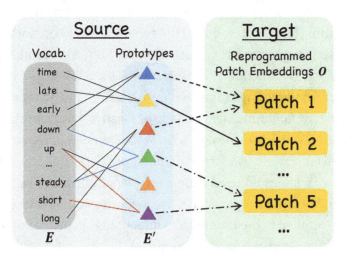

Figure 4-4. *Patch reprogramming [1](The above figure represents transforming the input patch to a language task.)*

Reprogramming essentially consists of adaptation and alignment. While we discussed the adaptation part so far, let us understand the alignment now. Refer to Figure 4-3. The translated patches are sent to a multi-head attention layer. From the previous chapter (Chapter 3) on transformers (where multi-head attention was explained), we know that this step helps in focusing on the different patches simultaneously. Each "head" processes information from the patches independently. This helps the model to capture various relationships and dependencies present in the data. Next, the processed data needs to be aligned with the specific format that the foundation model can use. The "linear projection" step helps to transform the dimensionality of the data (reprogrammed patches) to match the expected input size or format of the foundation model.

This ensures that the data can be correctly interpreted and processed by the model.

At this stage, the input to the foundation model is natural language. So similar to leveraging explanations of the task in prompts while using an LLM to get better output, we can add prompts to the patches. This prompt prefix complements patch programming to guide the LLM for better forecasts.

There are three main parts in the prompt for leveraging prompt as a prefix. We should pass (1) the dataset context, (2) task instruction, and (3) input statistics as part of the prompt. In the prompt example in Figure 4-5, the dataset context provides the LLM with essential background information about the input time series, which often exhibits distinct characteristics across various domains. Task instruction serves as a guide for the foundation model in the transformation of patch embeddings for specific tasks. The input time series is also enriched with additional statistics. These stats could be information regarding trends and lags which helps in pattern recognition and reasoning.

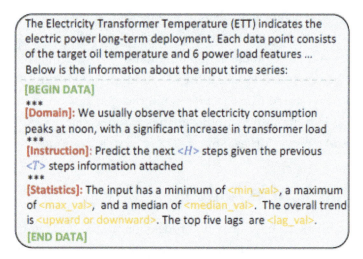

Figure 4-5. *Prompt as a prefix. <H/T> as task instruction and <min_val, max_val ..> as input statistics. [1]*

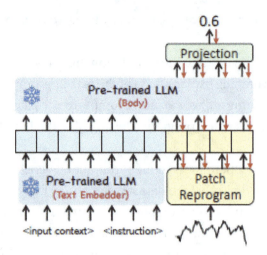

Figure 4-6. Prompt as a prefix and patches are sent to the foundation model [1]

The prompt and patches are fed to the LLM, and as a result the LLM generates output patch embeddings. Remember that the output is from a language-trained LLM, which needs to be converted to time series forecasts. This is achieved by flattening, and then linear projections are applied to get the desired output which is the time series forecast.

Before we get into practical implementation, readers may please note that what we understood so far is the idea and working behind Time-LLM. Please note that the practical implementations may differ and use clever tricks like computing input statistics automatically as part of the code and using FFT (Fast Fourier Transform) to compute lags.

4.2 Time-LLM in Action

Having established a high-level theoretical foundation of Time-LLM, we shall now translate abstract concepts into practical code implementation.

4.2.1 Univariate Use Case

Let's explore Time-LLM to solve a univariate problem.

Import required modules:

```
import torch
import psutil
import platform
import numpy as np
import pandas as pd
import matplotlib.pyplot as plt
from sklearn.metrics import mean_squared_error, mean_absolute_
percentage_error, r2_score
from neuralforecast import NeuralForecast
from neuralforecast.models import TimeLLM
from neuralforecast.utils import AirPassengersPanel, augment_
calendar_df
from transformers import GPT2Config, GPT2Model, GPT2Tokenizer
```

Let's display the GPU and CPU information:

```
use_cuda = torch.cuda.is_available()
if use_cuda:
    print('__CUDNN VERSION:', torch.backends.cudnn.version())
    print('__Number CUDA Devices:', torch.cuda.device_count())
    print('__CUDA Device Name:',torch.cuda.get_device_name(0))
    print('__CUDA Device Total Memory [GB]:',torch.cuda.get_
    device_properties(0).total_memory/1e9)


    __CUDNN VERSION: 90100
    __Number CUDA Devices: 1
    __CUDA Device Name: Tesla T4
    __CUDA Device Total Memory [GB]: 15.835660288
```

Let's print the memory, CPU, and platform information:

```python
mem = psutil.virtual_memory()

print("Available Memory:")
print("  Total:", mem.total / (1024 ** 2), "MB")
cpu_count = psutil.cpu_count()
cpu_count_logical = psutil.cpu_count(logical=True)

print("\nCPU Details:")
print("  Physical Cores:", cpu_count)
print("  Logical Cores:", cpu_count_logical)

platform_info = platform.platform()
print("\nPlatform:", platform_info)
```

```
        Available Memory:
          Total: 12978.96484375 MB

        CPU Details:
          Physical Cores: 2
          Logical Cores: 2

        Platform: Linux-6.1.85+-x86_64-with-glibc2.35
```

Let's load the AirPassenger dataset and split data into train and test:

```python
from neuralforecast.utils import AirPassengersDF
Y_df = AirPassengersDF
Y_df = Y_df.reset_index(drop=True)
Y_df.head()
```

	unique_id	ds	y
0	1.0	1949-01-31	112.0
1	1.0	1949-02-28	118.0
2	1.0	1949-03-31	132.0
3	1.0	1949-04-30	129.0
4	1.0	1949-05-31	121.0

```
Y_train_df = Y_df[Y_df.ds<='1959-12-31'] # 132 train
Y_test_df = Y_df[Y_df.ds>'1959-12-31']   # 12 test
```

Next, let's work on setting up GPT2:

```
gpt2_config = GPT2Config.from_pretrained('openai-
community/gpt2')
gpt2 = GPT2Model.from_pretrained('openai-community/gpt2',
config=gpt2_config)
gpt2_tokenizer = GPT2Tokenizer.from_pretrained('openai-
community/gpt2')
prompt_prefix = "The dataset contains data on monthly air
passengers. There is a yearly seasonality"
```

Let's initialize, train the model (TimeLLM), and define its hyperparameters.

h is the forecast horizon.

input_size is the autoregressive input size.

llm is the LLM model to be used.

llm_config is the configuration of LLM.

llm_tokenizer is the tokenizer of LLM.

prompt_prefix is the prompt to inform the LLM about the dataset.

batch_size is the number of different series in each batch.

windows_batch_size is the number of windows to sample in each training batch.

```
horizon = 12
timellm = TimeLLM(h=horizon,
                  input_size=36,
                  llm=gpt2,
                  llm_config=gpt2_config,
                  llm_tokenizer=gpt2_tokenizer,
                  prompt_prefix=prompt_prefix,
                  batch_size=24,
                  windows_batch_size=24)

nf = NeuralForecast(
    models=[timellm],
    freq='M'
)

nf.fit(df=Y_train_df, val_size=12)
```

Now that we completed training, let's work on prediction.

Note   It took 26 minutes to train the model and 25 minutes to generate predictions using the hardware specification printed above.

```
forecasts = nf.predict()
forecasts.head()
```

	ds	TimeLLM
unique_id		
1.0	1960-01-31	354.034912
1.0	1960-02-29	448.474457
1.0	1960-03-31	530.106445
1.0	1960-04-30	350.741852
1.0	1960-05-31	450.240967
1.0	1960-06-30	407.927460
1.0	1960-07-31	458.495087
1.0	1960-08-31	470.798126
1.0	1960-09-30	481.445496
1.0	1960-10-31	328.507050
1.0	1960-11-30	377.881073
1.0	1960-12-31	375.662720

Let's measure the model's accuracy:

```
calculate_error_metrics(Y_test_df[['y']],forecasts['TimeLLM'])
```

```
MSE : 9581.9990234375
RMSE : 97.88768768310547
MAPE : 0.172757089138031
r2 : -0.72976753530400161
adjusted_r2 : -0.9027442888344175
```

Let's visualize the predictions:

```
Y_train_df.set_index('ds',inplace =True)
forecasts.set_index('ds',inplace =True)
Y_test_df.set_index('ds',inplace =True)
plt.figure(figsize=(20, 3))
```

```
y_past = Y_train_df["y"]
y_pred = forecasts['TimeLLM']
y_test = Y_test_df["y"]
plt.plot(y_past, label="Past time series values")
plt.plot(y_pred, label="Forecast")
plt.plot(y_test, label="Actual time series values")
plt.title('AirPassengers Forecast', fontsize=10)
plt.ylabel('Monthly Passengers', fontsize=10)
plt.xlabel('Timestamp [t]', fontsize=10)

plt.legend();
```

Figure 4-7 helps us to appreciate that the air passenger count predicted by our model is not close to reality. Please refer to the "Summary" section for more details.

Figure 4-7. *Actual vs. predicted plot*

4.2.2 Multivariate Use Case

Now that we have tried the univariate use case, let's explore Time-LLM for the multivariate problem.

Import required modules:

```
import torch
import psutil
import platform
```

```python
import numpy as np
import pandas as pd
import pytorch_lightning as pl
import matplotlib.pyplot as plt
from sklearn.metrics import mean_squared_error, mean_absolute_
percentage_error, r2_score
from neuralforecast import NeuralForecast
from neuralforecast.models import TimeLLM
from neuralforecast.utils import AirPassengersPanel, augment_
calendar_df
from transformers import GPT2Config, GPT2Model, GPT2Tokenizer
```

Let's display the GPU and CPU information:

```python
use_cuda = torch.cuda.is_available()
if use_cuda:
    print('__CUDNN VERSION:', torch.backends.cudnn.version())
    print('__Number CUDA Devices:', torch.cuda.device_count())
    print('__CUDA Device Name:',torch.cuda.get_device_name(0))
    print('__CUDA Device Total Memory [GB]:',torch.cuda.get_
    device_properties(0).total_memory/1e9)
```

```
__CUDNN VERSION: 8902
__Number CUDA Devices: 1
__CUDA Device Name: Tesla T4
__CUDA Device Total Memory [GB]: 15.835660288
```

Let's print the memory, CPU, and platform information:

```python
mem = psutil.virtual_memory()
print("Available Memory:")
print("  Total:", mem.total / (1024 ** 2), "MB")
cpu_count = psutil.cpu_count()
```

```
cpu_count_logical = psutil.cpu_count(logical=True)
print("\nCPU Details:")
print("  Physical Cores:", cpu_count)
print("  Logical Cores:", cpu_count_logical)
platform_info = platform.platform()
print("\nPlatform:", platform_info)
```

```
        Available Memory:
          Total: 12978.96484375 MB

        CPU Details:
          Physical Cores: 2
          Logical Cores: 2

        Platform: Linux-6.1.85+-x86_64-with-glibc2.35
```

Let's load the AirPassenger dataset and split data into train and test:

```
AirPassengersPanel, calendar_cols = augment_calendar_
df(df=AirPassengersPanel, freq='M')
```

```
Y_train_df = AirPassengersPanel[AirPassengersPanel.ds<AirPassen
gersPanel['ds'].values[-12]]
Y_test_df = AirPassengersPanel[AirPassengersPanel.ds>=AirPassen
gersPanel['ds'].values[-12]].reset_index(drop=True)
```

Note that other features apart from "y" are added as exogenous variables:

```
Y_train_df.head()
```

	unique_id	ds	y	trend	y_[lag12]	month
0	Airline1	1949-01-31	112.0	0	112.0	-0.500000
1	Airline1	1949-02-28	118.0	1	118.0	-0.409091
2	Airline1	1949-03-31	132.0	2	132.0	-0.318182
3	Airline1	1949-04-30	129.0	3	129.0	-0.227273
4	Airline1	1949-05-31	121.0	4	121.0	-0.136364
...	...	...	...	...	...	...
271	Airline2	1959-08-31	859.0	271	805.0	0.136364
272	Airline2	1959-09-30	763.0	272	704.0	0.227273
273	Airline2	1959-10-31	707.0	273	659.0	0.318182
274	Airline2	1959-11-30	662.0	274	610.0	0.409091
275	Airline2	1959-12-31	705.0	275	637.0	0.500000

264 rows × 6 columns

Now let's work on setting up GPT2:

```
gpt2_config = GPT2Config.from_pretrained('openai-
community/gpt2')
gpt2 = GPT2Model.from_pretrained('openai-community/gpt2',
config=gpt2_config)
gpt2_tokenizer = GPT2Tokenizer.from_pretrained('openai-
community/gpt2')
prompt_prefix = "The dataset contains data on monthly air
passengers. There is a yearly seasonality"
```

Let's initialize, train the model (TimeLLM), and define its hyperparameters.

h is the forecast horizon.

input_size is the autoregressive input size.

llm is the LLM model to be used.

llm_config is the configuration of LLM.

llm_tokenizer is the tokenizer of LLM.

prompt_prefix is the prompt to inform the LLM about the dataset.

batch_size is the number of different series in each batch.

windows_batch_size is the number of windows to sample in each training batch.

```
horizon = 12
timellm = TimeLLM(h=horizon,
                  input_size=36,
                  llm=gpt2,
                  llm_config=gpt2_config,
                  llm_tokenizer=gpt2_tokenizer,
                  prompt_prefix=prompt_prefix,
                  batch_size=24,
                  windows_batch_size=24)

nf = NeuralForecast(
    models=[timellm],
    freq='M'
)

nf.fit(df=Y_train_df, val_size=12)
```

Now that training is done, let's try predicting by passing future exogenous which are part of test data prepared earlier.

Note   It took 26 minutes to train the model and 25 minutes to generate predictions using the hardware specification printed above.

Pass the Y_test_df which contains future exogenous variables:

```
forecasts = nf.predict(futr_df=Y_test_df)
```

Let's measure the model's accuracy:

```
calculate_error_metrics(Y_test_df[['y']],forecasts['TimeLLM'])
```

```
MSE : 9568.9033203125
RMSE : 97.82077026367188
MAPE : 0.13793222606182098
r2 : 0.6587345516012689
adjusted_r2 : 0.643222485764963
```

Let's see predictions for Airline1; TimeLLM is the predicted column:

```
print(forecasts['TimeLLM'][:12])
```

	TimeLLM
unique_id	
Airline1	340.337128
Airline1	438.639801
Airline1	516.180298
Airline1	354.548615
Airline1	449.864777
Airline1	415.037964
Airline1	457.495026
Airline1	450.726074
Airline1	485.232635
Airline1	335.309113
Airline1	376.948853
Airline1	374.285309

dtype: float32

Let's see predictions for Airline2; TimeLLM is the predicted column:

```
print(forecasts['TimeLLM'][12:])
```

TimeLLM

unique_id

Airline2	639.903687
Airline2	739.086670
Airline2	817.289307
Airline2	654.157349
Airline2	750.472900
Airline2	715.044678
Airline2	758.234802
Airline2	751.383972
Airline2	786.069458
Airline2	634.785767
Airline2	676.737732
Airline2	674.005310

dtype: float32

Let's visualize the predictions:

```
train_df_1 = Y_train_df[Y_train_df.unique_id == 'Airline1']
airline_df_1 = Y_test_df[Y_test_df.unique_id == 'Airline1']
train_df_1.set_index('ds',inplace =True)
forecasts.set_index('ds',inplace =True)
```

```
airline_df_1.set_index('ds',inplace =True)
plt.figure(figsize=(20, 3))
y_past = train_df_1["y"]
y_pred = forecasts['TimeLLM'][:12]
y_test = airline_df_1["y"]
plt.plot(y_past, label="Past time series values")
plt.plot(y_pred, label="Forecast")
plt.plot(y_test, label="Actual time series values")
plt.title('AirPassengers Forecast for Airline1', fontsize=10)
plt.ylabel('Monthly Passengers', fontsize=10)
plt.xlabel('Timestamp [t]', fontsize=10)

plt.legend();
```

Figure 4-8. *Actual vs. predicted plot*

Figure 4-8 helps us to appreciate that the air passenger count predicted by our model is not close to reality. Please refer to the "Summary" section for more details.

```
train_df_2 = Y_train_df[Y_train_df.unique_id == 'Airline2']
airline_df_2 = Y_test_df[Y_test_df.unique_id == 'Airline2']
train_df_2.set_index('ds',inplace =True)
airline_df_2.set_index('ds',inplace =True)
plt.figure(figsize=(20, 3))
y_past = train_df_2["y"]
y_pred = forecasts['TimeLLM'][12:]
y_test = airline_df_2["y"]
```

```
plt.plot(y_past, label="Past time series values")
plt.plot(y_pred, label="Forecast")
plt.plot(y_test, label="Actual time series values")
plt.title('AirPassengers Forecast for Airline2', fontsize=10)
plt.ylabel('Monthly Passengers', fontsize=10)
plt.xlabel('Timestamp [t]', fontsize=10)
plt.legend();
```

Figure 4-9 helps us to appreciate that the air passenger count predicted by our model is not close to reality. One of the reasons could be using such large models on smaller datasets, leading to over- or underfitting. Sometimes, these reprogrammed models are too complex for the available training set.

Figure 4-9. *Actual vs. predicted plot*

4.3 Summary

We understood how the Time-LLM repurposes a foundation model that is designed for NLP tasks and uses it for time series forecasting. We used patching to capture the LSI and make the best use of the underlying transformer/attention mechanism.

Finally, we learned how to do forecasting in both univariate and multivariate scenarios and understood how such large models might over- or underfit for smaller datasets. This model does not work well on datasets with strong temporal dynamics (changes in time series characteristics over time). In the next chapter, let's understand Chronos along with a sample implementation.

4.4 Reference

[1].   Time-LLM: Time Series Forecasting by Reprogramming Large Language Models; Ming Jin et.al. `https://doi.org/10.48550/arXiv.2310.01728`

CHAPTER 5

Chronos: Pre-trained Probabilistic Time Series Model

Chapter Goal: Learn how to leverage Chronos, a pre-trained probabilistic time series model.

5 Introduction

In the previous chapter, we understood how large language models are reprogrammed for time series forecasting.

The emergence of large language models with zero-shot learning capabilities has encouraged the development of foundation models for time series by directly using pre-trained LLMs in natural language and fine-tuning LLMs to handle time series tasks.

Zero-shot forecasting is the ability of models to generate forecasts for time series from unseen datasets. One of the popular techniques is training on a single time series dataset and testing on a different dataset.

Several methods adapting LLMs to the time series domain have been developed. One line of work treats numerical time series data as raw text and directly uses the pre-trained LLMs with minimal or no fine-tuning to

© Banglore Vijay Kumar Vishwas and Sri Ram Macharla 2025
B. V. Vishwas and Sri Ram Macharla, *Time Series Forecasting Using Generative AI*,
https://doi.org/10.1007/979-8-8688-1276-7_5

forecast unseen time series. LLM-based forecasting models such as Time-LLM have shown evidence that pre-trained models perform well at a zero-shot forecasting ability.

Chronos is a probabilistic pre-trained time series forecasting model based on T5 family language model architectures. It leverages existing language model architectures as both language and time series are sequential. The only difference is that their representation of natural language consists of words from a finite vocabulary, while time series are real valued.

5.1 Technical Overview of Chronos

Chronos is a language modeling framework minimally adapted for time series forecasting. Chronos tokenizes time series into discrete bins through simple scaling and quantization of real values. By using this method, we can train off-the-shelf language models with no changes to the model architecture as depicted in Figure 5-1. A straightforward approach proves to be effective in addressing a broad range of time series problems with minimal modifications.

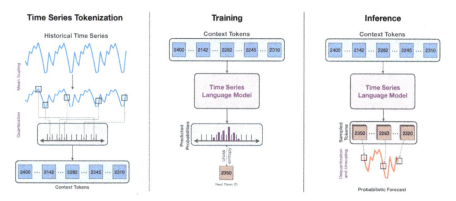

Figure 5-1. *(Left) Scaled and quantized input time series to obtain input from a sequence of tokens. (Center) Encoder-decoder or decoder-only model accepting tokens which is trained using cross-entropy loss. (Right) Multiple trajectories are sampled to obtain predictive distribution during inference by autoregressively sampling tokens from the model and mapping them to original numerical values [1]*

5.2 Time Series Tokenization

Time series data requires mapping the observations to a finite set of tokens, as originally language models operate on tokens from a finite vocabulary, to address this scale and then quantize observations into a fixed number of bins.

Scaling: The main goal of normalization is to transform the data to fit within a specific range suitable for quantization; there are several scaling techniques such as mean scaling, standard scaling, min-max scaling, MaxAbsScaler, RobustScaler, and several others as mean scaling is known to be effective in deep learning models commonly used for time series applications.

Quantization: This is a technique that converts real values into discrete tokens as scaled time series cannot be processed directly by language models. Uniform binning is used as it selects bin centers uniformly within some interval and as the distribution of training and inference data significantly differ, or any other quantization technique can also be used.

Time series tokens will also include two special tokens, PAD and EOS, which are commonly used in language models. The PAD token is used to pad time series of different lengths to a fixed length for batch construction and replace missing values. The EOS token is appended to the quantized and padded time series to denote the end of the sequence and also helps make training and inference of the large language models much easier.

5.3 Training

Tokenized time series are used to train and minimize the cross-entropy between the distribution of the quantized ground truth label and the predicted distribution.

Categorical cross-entropy loss is not a distance-aware objective function, which means it does not explicitly recognize that bin i is closer to bin i+1 than to i+2. Based on the distribution of bin indices in the training dataset, the models associated nearby bins together, which means Chronos performs regression via classification. This is unlike typical probabilistic time series forecasting models, which either use parametric continuous distributions such as Gaussian and Student's t-distribution or perform quantile regression.

The benefits of using categorical outputs are that we can use existing language models with no modification to the architecture or training objective, and we can use them out of the box, and they don't impose any restrictions on the structure of the output distribution.

5.4 Inference

Context tokens are fed into the model to generate the future tokens; then these tokens need to be mapped back to real values and then unscaled to obtain original forecasts. The dequantization function is used to map the predicted tokens to real values and then unscaled by applying the inverse scaling transformation.

Training these models requires a large volume of data, and public time series data is barely available, which poses challenges in training zero-shot forecasting models. To tackle this, we can diversify the training data by generating mix-up data augmentation for real datasets and using synthetic data for training, which can be done using techniques such as TSMixup (time series mix-up) and KeralSynth (synthetic data generation using the Gaussian process).

5.5 Chronos in Action

Having established a high-level theoretical foundation of Chronos, we shall now translate abstract concepts into practical code implementation of **Chronos-tiny**.

5.5.1 Chronos-tiny Use Case

At the time of book development, the Chronos forecasting scope was limited to univariate forecasting.

Import required modules:

```
import autogluon
import numpy as np
from sklearn.metrics import mean_squared_error, mean_absolute_
percentage_error, r2_score
```

```
from autogluon.timeseries import TimeSeriesPredictor,
TimeSeriesDataFrame
from autogluon.timeseries.models import WaveNetModel
import pandas as pd
```

Let's load the AirPassengersDataset CSV using pandas, a dataset which contains 12 years of monthly air passenger data:

```
Y_df = pd.read_csv('AirPassengersDataset.csv')
Y_df = Y_df.reset_index(drop=True)
Y_df.head()
```

	item_id	timestamp	target
0	airline_1	1949-01-31	112
1	airline_1	1949-02-28	118
2	airline_1	1949-03-31	132
3	airline_1	1949-04-30	129
4	airline_1	1949-05-31	121

AutoGluon expects time series data in long format. Each row of the data frame contains a single observation (time step) of a single time series represented by

a) Unique ID of the time series item_id as int or str

b) Timestamp of the observation timestamp as a pandas.Timestamp or compatible format

c) Numeric value of the time series target

```python
Y_df['ds'] = pd.to_datetime(Y_df['ds'])
Y_df.rename(columns={"ds":"timestamp","unique_id" :"item_id",
"y": "target"},inplace = True)
Y_df['item_id'] = 'airline_1'

data = TimeSeriesDataFrame.from_data_frame(
    Y_df,
    id_column="item_id",
    timestamp_column="timestamp"
)
data.tail()
```

		target
item_id	timestamp	
airline_1	1960-08-31	606
	1960-09-30	508
	1960-10-31	461
	1960-11-30	390
	1960-12-31	432

Split data into train and test. Separate the last 1-year data for the test and use the remaining 11 years of data to train and predict:

```python
train_data = data.head(132)
test_data = data.tail(12)
```

Create a TimeSeriesPredictor object to forecast future values and explicitly define chronos_tiny to be used:

```
prediction_length =12
predictor = TimeSeriesPredictor(prediction_length=prediction_
length).fit(
train_data, presets="chronos_tiny"
)
```

Predict the next defined horizon:

```
predictions = predictor.predict(train_data)
predictions.head()
```

The predicted column of interest is mean.

item_id	timestamp	mean	0.1	0.2	0.3	0.4	0.5	0.6	0.7	0.8	0.9
airline_1	1960-01-31	409.422791	384.886261	388.735138	391.429330	396.047955	412.790543	415.677185	422.412680	431.457489	441.272122
	1960-02-29	418.852448	398.357269	401.436371	406.247467	413.367871	419.526031	422.990009	426.453979	431.842377	443.004117
	1960-03-31	434.440369	409.518991	415.292297	418.948700	424.914441	430.110397	436.845886	448.969839	454.165796	464.750153
	1960-04-30	439.636383	396.240417	405.670105	419.141153	432.612164	441.657013	451.856476	454.165802	458.784424	464.365262
	1960-05-31	468.791504	426.838867	434.151703	448.777393	454.165802	458.014664	482.647388	487.458466	496.348627	502.468994
	1960-06-30	533.548645	475.142120	498.812579	507.664978	511.898743	522.483093	531.143042	550.387329	577.329407	586.374200

Measure the model's accuracy:

```
calculate_error_metrics(test_data['target'],
predictions['mean']['airline_1'])
```

```
MSE : 692.7698855972849
RMSE : 26.320522137626465
MAPE : 0.04469724962539968
r2 : 0.874939370068838
adjusted_r2 : 0.8624333070757219
```

Visualize the predictions:

```
predictor.plot(
    data=Y_df,
    predictions=predictions,
    item_ids=["airline_1"],
    max_history_length=200,
);
```

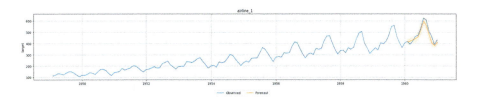

Figure 5-2. *Observed vs. forecast*

Figure 5-2 helps us to appreciate that the air passenger count predicted by our model is close to reality.

5.5.2 chronos_large_ensemble Use Case

Let's now implement **chronos_large_ensemble** using the same dataset as above which builds an ensemble of seasonal naive, tree-based, and deep learning models with fast inference.

Import required modules:

```
import autogluon
import numpy as np
from sklearn.metrics import mean_squared_error, mean_absolute_
percentage_error, r2_score
from autogluon.timeseries import TimeSeriesPredictor,
TimeSeriesDataFrame
import pandas as pd
```

Let's load the AirPassengersDataset CSV using pandas, a dataset that contains 12 years of monthly air passenger data:

```
Y_df = pd.read_csv('AirPassengersDataset.csv')
Y_df = Y_df.reset_index(drop=True)
Y_df.head()
```

	item_id	timestamp	target
0	airline_1	1949-01-31	112
1	airline_1	1949-02-28	118
2	airline_1	1949-03-31	132
3	airline_1	1949-04-30	129
4	airline_1	1949-05-31	121

AutoGluon expects time series data in long format. Each row of the data frame contains a single observation (time step) of a single time series represented by

a) Unique ID of the time series item_id as int or str

b) Timestamp of the observation timestamp as a pandas.Timestamp or compatible format

c) Numeric value of the time series target

```
Y_df['ds'] = pd.to_datetime(Y_df['ds'])
Y_df.rename(columns={"ds":"timestamp","unique_id" :"item_id",
"y": "target"},inplace = True)
Y_df['item_id'] = 'airline_1'
Y_df.head()
```

```
data = TimeSeriesDataFrame.from_data_frame(
    Y_df,
    id_column="item_id",
    timestamp_column="timestamp"
)
data.tail()
```

		target
item_id	timestamp	
airline_1	1960-08-31	606
	1960-09-30	508
	1960-10-31	461
	1960-11-30	390
	1960-12-31	432

Split data into train and test. Separate the last 1-year data for the test and use the remaining 11 years of data to train and predict:

```
train_data = data.head(132)
test_data = data.tail(12)
```

Create a TimeSeriesPredictor object to forecast future values and explicitly define chronos_large_ensemble to be used:

```
prediction_length =12
predictor = TimeSeriesPredictor(prediction_length=prediction_
length).fit(
train_data, presets="chronos_large_ensemble"
)
```

Predict the next defined horizon:

```
predictions = predictor.predict(train_data)
predictions.head()
```

The predicted column of interest is the mean.

item_id	timestamp	mean	0.1	0.2	0.3	0.4	0.5	0.6	0.7	0.8	0.9
airline_1	1960-01-31	409.050995	390.085909	396.596233	401.290635	405.301829	409.050995	412.800161	416.811354	421.505756	428.016081
	1960-02-29	388.893677	362.072995	371.279984	377.918871	383.591555	388.893677	394.195798	399.868482	406.507369	415.714359
	1960-03-31	453.047974	420.199481	431.475694	439.606637	446.554228	453.047974	459.541720	466.489311	474.620253	485.896466
	1960-04-30	444.603302	406.673130	419.693779	429.082583	437.104970	444.603302	452.101634	460.124021	469.512825	482.533474
	1960-05-31	467.213440	424.806218	439.363746	449.860749	458.830050	467.213440	475.596830	484.566131	495.063133	509.620662
	1960-06-30	528.337036	481.882252	497.829225	509.328115	519.153492	528.337036	537.520580	547.345957	558.844847	574.791820

Measure the model's accuracy:

```
calculate_error_metrics(test_data['target'],predictions['mean']
['airline_1'])
```

```
MSE : 231.54919850407168
RMSE : 15.216740731972523
MAPE : 0.02696902830021881
r2 : 0.9582001336562018
adjusted_r2 : 0.954020147021822
```

Visualize the predictions:

```
predictor.plot(
    data=Y_df,
    predictions=predictions,
    item_ids=["airline_1"],
    max_history_length=200,
);
```

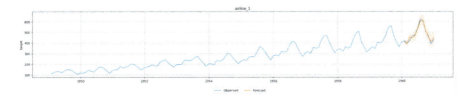

Figure 5-3. *Observed vs. forecast*

Figure 5-3 helps us to appreciate that the air passenger count predicted by our model is close to reality.

5.6 Summary

In this chapter, we understood how Chronos pre-trained probabilistic time series model works and how to implement chronos_tiny and chronos_large_ensemble using real-world datasets. Chronos can be used for any univariate forecasting use case, and it works best on datasets that consist of observations captured at equal intervals of time. In the next section, let's deep dive into TimeGPT.

5.7 Reference

[1].   Chronos: Learning the Language of Time Series.
https://doi.org/10.48550/arXiv.2403.07815

CHAPTER 6

TimeGPT: The First Foundation Model for Time Series

Chapter Goal: Learn how to leverage TimeGPT to build robust and accurate time series models.

6 Introduction

In the previous chapter, we explored the technical overview of Chronos, a pre-trained time series forecasting model, and completed hands-on implementation using a real-world dataset.

In this chapter, let's explore TimeGPT, a production-ready, generative pre-trained transformer for time series capable of predicting retail, electricity, web traffic, transport, economics, finance, and IoT with just a few lines of code. We can also understand why it is called the first foundation model for time series.

Time series analysis historically relied on traditional techniques like Fourier Analysis, Moving Average, Autoregressive, Autoregressive Integrated Moving Average, Exponential Smoothing, Vector Autoregression, Theta, and Generalized Autoregressive Conditional Heteroskedasticity (GARCH) Models, which were popular historically and

B. V. Vishwas and Sri Ram Macharla, *Time Series Forecasting Using Generative AI*, https://doi.org/10.1007/979-8-8688-1276-7_6

used in various domains, and later evolved to more powerful machine learning tools like Random Forest, Gradient Boosting, XGBoost, LightGBM, CatBoost, and Prophet.

The rapid advancement in computational power, availability and storage of large datasets, and advancements in algorithms and architectures have fueled the advent of new deep learning methodologies in some use cases to outperform traditional techniques. Deep learning is a global approach that offers advantages over conventional methods in automatic feature learning, handling large and complex data, improved performance, handling nonlinear relationships, handling structured and unstructured data, and handling sequential data.

Deep learning models such as RNN, LSTM, GRU, and CNN, designed for natural language processing and computer vision, when repurposed for sequential data demonstrated amazing capabilities in learning patterns. Significant advancements in hardware and distributed parallel processing fueled the popularity of transformer models, which have gained popularity in recent years as they demonstrate amazing capabilities for learning from large volumes of data.

Recent advancements in transformer architecture have led to the development of powerful transformer-based models like Autoformer, Informer, FEDformer, and PatchTST. These models leverage self-attention and innovative techniques to capture long-range dependencies and complex patterns in time series data. Further to the discussion so far, let us explore TimeGPT – the first time series foundation model.

TimeGPT was trained on a huge volume of publicly available datasets, collectively over 100 billion data points, using NVIDIA A10G GPU for multiple days. The training set included a wide range of domains due to this dataset's comprised of a diverse selection of temporal patterns, structural breaks, seasonality, cycles of various lengths, various trends, and irregular and regular patterns, offering a robust training dataset. This pre-training allows it to generalize well to new, unseen time series data, making it a powerful and versatile tool for time series forecasting and analysis.

The selection of diverse datasets helps TimeGPT to forecast unseen time series accurately while eliminating the need for individual model training and optimization and performs well on single-series and multiple-series forecasting as depicted in Figure 6-1.

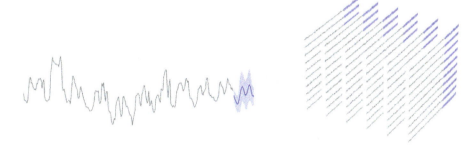

Figure 6-1. *(a) Single-series forecasting, (b) multiple-series forecasting [1]*

6.1 Technical Overview of TimeGPT

TimeGPT is a transformer-based model and employs an encoder-decoder architecture with multiple layers, each incorporating residual connections and layer normalization. The decoder's output is projected to the forecasting window dimension through a linear layer. Local positional encoding is added to the window of historical values to enhance the input. The attention mechanisms allow models to focus on the most relevant parts of the input sequence, improving their ability to capture long-range dependencies and make accurate predictions.

TimeGPT can handle different input sizes, horizons, and other characteristics within data such as frequency, sparsity, trend, seasonality, stationarity, and heteroskedasticity which may present distinct complications for both local and global models.

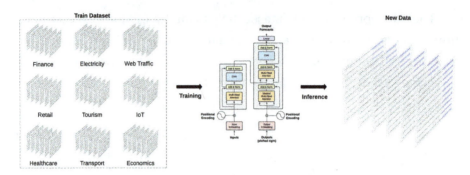

Figure 6-2. *Depicting datasets from different domains used for training and inference are generated on new data [1]*

The forecasting model is a function:

f_θ: $\mathcal{X} \to \mathcal{Y}$, where $\mathcal{X}$ is the feature space and $\mathcal{Y}$ is the dependent variable.

Consider the setting: $\mathcal{X} = \{y_{[0:t]}, x_{[0:t+h]}\}$ and $\mathcal{Y} = \{y_{[t+1:t+h]}\}$

where h is the horizon to forecast, y is the target variable, and x is the exogenous variable. The goal is to estimate the conditional distribution.

$$P\left(y_{[t+1:t+h]} \mid y_{[0:t]}, x_{[0:t+h]}\right) = f_\theta\left(y_{[0:t]}, x_{[0:t+h]}\right)$$

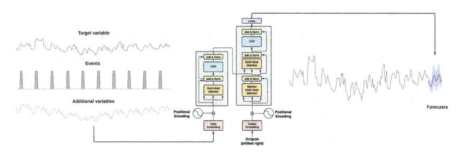

Figure 6-3. *Inference of new time series [1]*

TimeGPT makes predictions by reading input series like humans by looking at the windows of past data, which are similar to tokens, and predicts the next required horizon. TimeGPT performs remarkably well on zero-shot inference which is with no fine-tuning, which is remarkable when compared against statistical models and state-of-the-art deep learning approaches. A known drawback of this model is that accuracy diminishes when the forecasting horizon is too long.

It's worth noting that the original paper on TimeGPT [1] provides a high-level overview of the approach, but lacks granular details regarding the specific implementation techniques.

6.2 TimeGPT in Action

Having established a high-level theoretical foundation of TimeGPT, we shall now translate abstract concepts into practical code implementation.

6.2.1 Setting Up an API Key for TimeGPT

There are secured and unsecured methods of the API key configuration process; we follow the unsecured method in this example.

  a) Log in to the NIXTLA Developer Dashboard by authenticating using Gmail, GitHub, or email.

b) In the dashboard, navigate to API Keys and click
    Create New API Key.

c) Paste the key directly into your Python code, by
    instantiating the NixtlaClient with your API key:

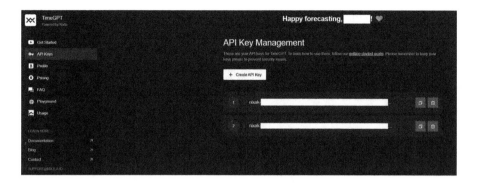

6.2.2 Univariate Use Case

Let's explore an example using a univariate dataset.

Import required libraries:

```
import numpy as np
import pandas as pd
from nixtla import NixtlaClient
import matplotlib.pyplot as plt
from sklearn.metrics import mean_squared_error, mean_absolute_
percentage_error, r2_score
```

Setting up your API key, copy and paste your key directly into your
Python code:

```
nixtla_client = NixtlaClient(
    api_key = 'nixtla-tok-xxxxxxxxxx'
)
```

Validate your API key, check the status of your API key, and use the
validate_api_key method of the NixtlaClient class. This method will return
True if the API key is valid and False otherwise.

```
nixtla_client.validate_api_key()
```

```
                              True
```

Load the AirPassenger dataset.

Split data into train and test. Separate the last 12 months of data for testing and use the remaining 11 years for model training:

```
Y_df = pd.read_csv('AirPassengersDataset.csv')
Y_df.drop(['unique_id'], axis =1 , inplace = True)
Y_df = Y_df.reset_index(drop=True)
Y_df.head()
```

	ds	y
0	1949-01-31	112
1	1949-02-28	118
2	1949-03-31	132
3	1949-04-30	129
4	1949-05-31	121

```
Y_train_df = Y_df[Y_df.ds<='1959-12-31']
Y_test_df = Y_df[Y_df.ds>'1959-12-31']
```

Plot the training data:

```
nixtla_client.plot(Y_train_df, time_col='ds', target_col='y')
```

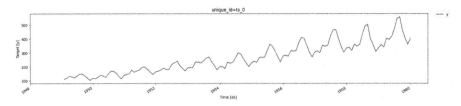

Figure 6-4. *Training data*

Set the following parameters:

a) df: A pandas DataFrame

b) h: Horizons

c) time_col: The column that identifies the date stamp

d) target_col: The forecast variable

```
timegpt_fcst_df = nixtla_client.forecast(df=Y_train_df, h=12,
time_col='ds', target_col='y')
timegpt_fcst_df.head()
```

	ds	TimeGPT
0	1960-01-31	414.514282
1	1960-02-29	403.244263
2	1960-03-31	443.127197
3	1960-04-30	439.750916
4	1960-05-31	462.836548

Check how well the model works on test data:

```
calculate_error_metrics(Y_test_df[['y']],timegpt_fcst_
df['TimeGPT'])
```

```
MSE : 213.93578505244417
RMSE : 14.62654385193044
MAPE : 0.026964465573293617
r2 : 0.9613797530757142
adjusted_r2 : 0.9575177283832856
```

Plot actual vs. predicted:

```
Y_train_df.set_index('ds',inplace =True)
timegpt_fcst_df.set_index('ds',inplace =True)
Y_test_df.set_index('ds',inplace =True)
plt.figure(figsize=(20, 5))
y_past = Y_train_df["y"][-50:]
y_pred = timegpt_fcst_df['TimeGPT']
y_test = Y_test_df["y"]
```

```
plt.plot(y_past, label="Past time series values")
plt.plot(timegpt_fcst_df, label="Forecast")
plt.plot(y_test, label="Actual time series values")
plt.title('AirPassengers Forecast', fontsize=10)
plt.ylabel('Monthly Passengers', fontsize=10)
plt.xlabel('Timestamp [t]', fontsize=10)
plt.tight_layout()
plt.xticks(rotation=90)
plt.legend();
```

Figure 6-5 helps us to appreciate that the air passenger count predicted by our model is close to reality.

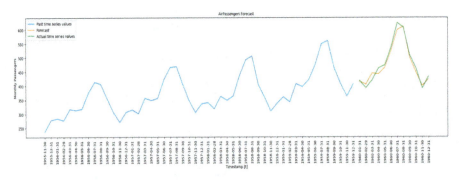

Figure 6-5. *Observed vs. forecast*

6.2.3 Multivariate Use Case

Let's explore an example using a multivariate dataset.

Import required libraries:

```
import numpy as np
import pandas as pd
from nixtla import NixtlaClient
import matplotlib.pyplot as plt
from sklearn.metrics import mean_squared_error, mean_absolute_
percentage_error, r2_score
```

Setting up your API key, copy and paste your key directly into your Python code:

```
nixtla_client = NixtlaClient(
    api_key = 'nixtla-tok-xxxxxxxxxxxx'
)
```

Validate your API key, check the status of your API key, and use the validate_api_key method of the NixtlaClient class. This method will return True if the API key is valid and False otherwise.

```
nixtla_client.validate_api_key()
```

True

Load the bike-sharing dataset [2]:

```
Y_df = pd.read_csv('Bike_sharing_systems.csv')
Y_df.drop(columns=['instant','casual','cnt'], axis =1,
inplace =True)
Y_df.rename(columns={'dteday':'ds','registered':'y'},
inplace=True)
Y_df.head()
```

	ds	season	yr	mnth	holiday	weekday	workingday	weathersit	temp	atemp	hum	windspeed	y
0	2011-01-01	1	0	1	0	6	0	2	0.344167	0.363625	0.805833	0.160446	654
1	2011-01-02	1	0	1	0	0	0	2	0.363478	0.353739	0.696087	0.248539	670
2	2011-01-03	1	0	1	0	1	1	1	0.196364	0.189405	0.437273	0.248309	1229
3	2011-01-04	1	0	1	0	2	1	1	0.200000	0.212122	0.590435	0.160296	1454
4	2011-01-05	1	0	1	0	3	1	1	0.226957	0.229270	0.436957	0.186900	1518

Split data into train and test:

```
Y_train_df = Y_df.iloc[:-24,:]
Y_test_df = Y_df.iloc[-24:,:]
```

Using the test data, let's remove the 'y' variable and make it a dataframe containing future exogenous variables for the defined horizon:

```
Y_test_df_w_y = Y_test_df.copy()
Y_test_df.drop(columns=['y'], axis =1, inplace =True)
Y_test_df_wo_y = Y_test_df
```

With the target variable:

```
Y_test_df_w_y.head()
```

	ds	season	yr	mnth	holiday	weekday	workingday	weathersit	temp	atemp	hum	windspeed	y
707	2012-12-08	4	1	12	0	6	0	2	0.381667	0.389508	0.911250	0.101379	4429
708	2012-12-09	4	1	12	0	0	0	2	0.384167	0.390146	0.905417	0.157975	2787
709	2012-12-10	4	1	12	0	1	1	2	0.435833	0.435575	0.925000	0.190308	4841
710	2012-12-11	4	1	12	0	2	1	2	0.353333	0.338363	0.596667	0.296037	5219
711	2012-12-12	4	1	12	0	3	1	2	0.297500	0.297338	0.538333	0.162937	5009

Without the target variable:

```
Y_test_df_wo_y.head()
```

	ds	season	yr	mnth	holiday	weekday	workingday	weathersit	temp	atemp	hum	windspeed
707	2012-12-08	4	1	12	0	6	0	2	0.381667	0.389508	0.911250	0.101379
708	2012-12-09	4	1	12	0	0	0	2	0.384167	0.390146	0.905417	0.157975
709	2012-12-10	4	1	12	0	1	1	2	0.435833	0.435575	0.925000	0.190308
710	2012-12-11	4	1	12	0	2	1	2	0.353333	0.338363	0.596667	0.296037
711	2012-12-12	4	1	12	0	3	1	2	0.297500	0.297338	0.538333	0.162937

Calling the forecast method and passing the exogenous variables:

```
timegpt_fcst_ex_vars_df = nixtla_client.forecast(df=Y_train_df,
X_df=Y_test_df_wo_y, h=24,)
timegpt_fcst_ex_vars_df.head()
```

	ds	TimeGPT
0	2012-12-08	3986.070068
1	2012-12-09	2798.701172
2	2012-12-10	4718.244629
3	2012-12-11	5315.483887
4	2012-12-12	5794.339844

Check how well the model predicts:

```
calculate_error_metrics(Y_test_df_w_y[['y']],timegpt_fcst_ex_
vars_df['TimeGPT'])
```

```
MSE : 2598206.886045963
RMSE : 1611.895432726938
MAPE : 0.9761389720503738
r2 : 0.05391699593098909
adjusted_r2 : 0.010913223018761298
```

Plot actual vs. predicted:

```
Y_train_df.set_index('ds',inplace =True)
timegpt_fcst_ex_vars_df.set_index('ds',inplace =True)
Y_test_df_w_y.set_index('ds',inplace =True)
plt.figure(figsize=(20, 5))
y_past = Y_train_df["y"][-75:]
y_pred = timegpt_fcst_ex_vars_df['TimeGPT']
y_test = Y_test_df_w_y["y"]
plt.plot(y_past, label="Past count of registered users")
plt.plot(y_pred, label="Forecast")
plt.plot(y_test, label="Actual count of registered users")
plt.title('Bike Sharing Forecast', fontsize=10)
plt.ylabel('Daily Count', fontsize=10)
plt.xlabel('Timestamp [t]', fontsize=10)
plt.xticks(rotation=90)
plt.legend();
```

Figure 6-6 helps us to appreciate that the count predicted by our model is not close to reality.

Figure 6-6. *Observed vs. forecast*

6.3 Summary

In this chapter, we understood how TimeGPT, the first foundation model for time series, works internally. We also tried a hands-on implementation using univariate and multivariate examples. In the next chapter, let's explore MOIRAI, a time series foundation model for universal forecasting.

6.4 References

[1].   TimeGPT-1. https://doi.org/10.48550/arXiv.2310.03589

[2].   Bike sharing dataset. https://doi.org/10.24432/C5W894

CHAPTER 7

MOIRAI: A Time Series LLM for Universal Forecasting

Chapter Goal: Learn how a universal forecasting model is designed and learn with real-world datasets.

7 Introduction

Ever wondered why we need so many models? Many a time working on use cases, you may have wondered why we develop a new model whenever the use case changes, while the task remains the same, for example, the classification of images or text for different domains. Another scenario could be time series forecasting for sales or crop yield. While the task of forecasting remains the same, we end up developing a new model every time. Some questions arise: Is it possible to reuse a model? Can we have a universal forecasting model? Let us understand how MOIRAI, a foundation model pre-trained on a large collection of time series datasets, tries to answer some of these questions.

© Banglore Vijay Kumar Vishwas and Sri Ram Macharla 2025
B. V. Vishwas and Sri Ram Macharla, *Time Series Forecasting Using Generative AI*,
https://doi.org/10.1007/979-8-8688-1276-7_7

The **M**asked Enc**O**der-based Un**I**ve**R**s**Al** Time Series Forecasting Transformer, or in short MOIRAI, is a result of novel enhancements made to the traditional time series forecasting transformer. The current version of the model was developed by training on a dataset containing observations taken across nine domains and has around 27B observations. In our experience, we found MOIRAI useful in scenarios involving zero-shot forecasting. In some use cases, it was providing results on par with multi-shot forecasting models. MOIRAI is trained and made available in three sizes, MOIRAISmall, MOIRAIBase, and MOIRAILarge, with 14m, 91m, and 311m parameters, respectively.

Zero-shot time series forecasting is the ability of the forecasting model to provide predictions on unseen datasets. It provides predictions without explicit training on past data for that specific use case. The model leverages its knowledge and patterns learned from pre-training from similar or related data. During one of the interactions with a potential client who was planning to release a new product, we were asked – how would you forecast the future sales of this product, as it has no historical data? Our answer, as you might have guessed by now, is zero-shot forecasting.

7.1 Challenges with Building a Universal Forecasting Model

To build a universal time series forecasting model, the unique challenges inherent to time series datasets need to be addressed. Challenges like (i) cross-frequency learning, (ii) accommodating an arbitrary number of variates for multivariate time series, and (iii) addressing the varying distributional properties inherent in large-scale data are handled by making some enhancements to the conventional transformer architecture. Let us understand these challenges in detail.

The frequency at which the observations in the time series dataset are gathered has a profound impact on the patterns in the dataset. To commence, the temporal granularity be it hourly, daily, or monthly intervals of a time series is profoundly consequential in delineating the intricate patterns that emerge within its temporal tapestry. To understand this better, think of the patterns present in a time series dataset having the numbers related to hourly, daily, weekly, and annual sales of a product, sold in the last five years. To make it still easier, think of withdrawing money from your savings account periodically on a daily, weekly, or monthly basis. Cross-frequency learning or learning with time series datasets sampled across different frequencies helps to improve forecasting results. The information learned from various frequencies helps to better understand the latent patterns. However, there are challenges related to overfitting, negative interference, and computational complexity. Negative interference is nothing but the degradation in models' performance across different frequencies of datasets.

Time series datasets are inherently heterogeneous, as there can be variation in several variables recorded at any point in time. Consider univariate and multivariate time series data related to sales of a product. The univariate primarily has only a value reflecting the number of units sold. The multivariate, additional could have values related to profit and color of the unit sold. The universal model developed should be flexible to handle multivariate interactions and exogenous covariates.

Probabilistic forecasting (recollect the DeepAR model) has its significance in time series forecasting. This is because it provides a comprehensive view of uncertainty and a range of possible forecasts rather than a single forecast. A multitude of time series datasets have differing statistical and probabilistic distributions. Using a Gaussian distribution as the predictive distribution has many benefits like mathematical simplicity (has only mean and variance), uncertainty quantification, and flexibility with additive noise models. However, it is not suitable for time series data with all observations greater than zero (positive time series). This makes it challenging to use the standard approach of using a simple distribution across a wide variety of datasets.

Any pre-trained model intended for universal forecasting necessitates training on large datasets from various domains. Existing generally available datasets are incapable of enabling the development of a universal forecasting model.

7.2 Technical Overview of MOIRAI

Now that we understand the challenges of developing a universal forecasting model, let us understand the architecture and approaches taken in developing MOIRAI to overcome the challenges.

MOIRAI leverages patches and masked encoder architecture to model time series. Parts of the input data are selectively hidden using a mask. The model is encouraged to predict the masked portions using known (unmasked) data. The model learns better by utilizing contextual understanding by predicting the masked data. The technique enhances learning of dependencies and improves forecasting accuracy.

The challenge of a universal forecaster (MOIRAI) to cater to multiple datasets with varying frequencies is dealt with the help of a layer containing multiple patches of sizes. Referring to Figure 7-1, we can see the layer with varying patch sizes (multi-patch size). MOIRAI uses a strategy wherein high-frequency data are handled with a larger patch size and low-frequency data are handled with a smaller patch size. This reduces computational complexity while maintaining long context length for high-frequency data. The benefit of this flexibility helps to transfer computation to transformer layers instead of embedding layers while dealing with low-frequency datasets. This makes the best use of computational time and resources too.

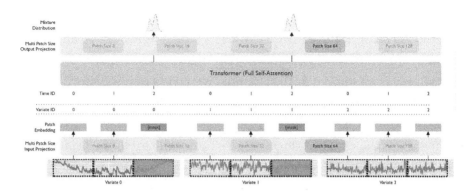

Figure 7-1. *The diagram represents the overall architecture of MOIRAI*

[In Figure 7-1, variate 0 and variate 1 are target variables, while variate 2 is the dynamic covariate. Considering a patch of size 64, the variates are transformed into patches of three tokens. That is, each 64-element patch is represented by three tokens, encapsulating key features of the patch. These tokens are then converted to patch embeddings (high-dimensional vectors) that represent the semantic meaning of each patch. These patch embeddings along with the sequence number (position of the overall data) and variate ID (which indicates a specific variable of the data) are input to the transformer. The patches (shaded) in the multi-patch size output projection layer represent the forecast horizon. The respective output representations from this layer are mapped to mixture distributions.]

The next challenge for a universal forecaster (MOIRAI) is to cater to multiple datasets with an arbitrary number of variates. This is addressed by using a novel approach called any-variate attention and making use of binary attention biases. The any-variate attention handles an arbitrary number of variates, while binary attention biases help to differentiate and encode the indexes (positions) of the variates. All variates are taken as a single sequence by flattening the multivariate time series data. The variate encodings help to distinguish between multiple variates in the data.

187

The next challenge of the foundation model to perform probabilistic forecasting using a simple distribution is handled by MOIRAI using a mixture of parametric distributions. The mixture comprises the following distributions: (i) student's t-distribution, which is a robust distribution option for time series; (ii) negative binomial distribution for positive count data; (iii) log-normal distribution, proven to be useful for scenarios with right-skewed data; and (iv) low variance normal distribution, useful for high confidence predictions.

Coming to the limitations, it is observed that forecasting results for use cases involving high-dimensional datasets are not accurate.

7.3 MOIRAI in Action

Having established a high-level theoretical foundation of MOIRAI, we shall now translate abstract concepts into practical code implementation of MOIRAI-Small.

Import required modules:

```
import torch
import numpy as np
import pandas as pd
import matplotlib.pyplot as plt
import matplotlib.pyplot as plt
import pandas as pd
import numpy as np
from einops import rearrange
from gluonts.dataset.multivariate_grouper import
MultivariateGrouper
```

```python
from gluonts.dataset.pandas import PandasDataset
from gluonts.dataset.split import split
from uni2ts.eval_util.plot import plot_single, plot_next_multi
from uni2ts.model.moirai import MoiraiForecast, MoiraiModule
from sklearn.metrics import mean_squared_error, mean_absolute_
percentage_error, r2_score
import warnings
warnings.filterwarnings('ignore')
```

Let's load the AirPassengers CSV using pandas, a dataset that contains 12 years of monthly air passenger data:

```python
df = pd.read_csv('AirPassengersDataset.csv')
df.rename(columns={'y': 'target'}, inplace=True)
df.drop(columns=['unique_id'], inplace=True)
df["ds"] = pd.to_datetime(df["ds"])
df.set_index("ds", inplace=True)
print(f"total length: {df.shape[0]}")
print(f"time frequency: {df.index.diff()[1]}")

df
```

ds	target
1949-01-31	112
1949-02-28	118
1949-03-31	132
1949-04-30	129
1949-05-31	121
...	...
1960-08-31	606
1960-09-30	508
1960-10-31	461
1960-11-30	390
1960-12-31	432

144 rows × 1 columns

Let's create a sample data with monthly frequency and split the data into train and test, which are input and labels:

```
inp = {
    "target": df["target"].to_numpy()[:120],
    "start": df.index[0].to_period(freq="M"),
}
label = {
    "target": df["target"].to_numpy()[120:144],
    "start": df.index[120].to_period(freq="M"),
}
```

Inp

```
{'target': array([112, 118, 132, 129, 121, 135, 148, 148, 136, 119, 104, 118, 115,
        126, 141, 135, 125, 149, 170, 170, 158, 133, 114, 140, 145, 150,
        178, 163, 172, 178, 199, 199, 184, 162, 146, 166, 171, 180, 193,
        181, 183, 218, 230, 242, 209, 191, 172, 194, 196, 196, 236, 235,
        229, 243, 264, 272, 237, 211, 180, 201, 204, 188, 235, 227, 234,
        264, 302, 293, 259, 229, 203, 229, 242, 233, 267, 269, 270, 315,
        364, 347, 312, 274, 237, 278, 284, 277, 317, 313, 318, 374, 413,
        405, 355, 306, 271, 306, 315, 301, 356, 348, 355, 422, 465, 467,
        404, 347, 305, 336, 340, 318, 362, 348, 363, 435, 491, 505, 404,
        359, 310, 337]),
 'start': Period('1949-01', 'M')}
```

Label

```
{'target': array([360, 342, 406, 396, 420, 472, 548, 559, 463, 407, 362, 405, 417,
        391, 419, 461, 472, 535, 622, 606, 508, 461, 390, 432]),
 'start': Period('1959-01', 'M')}
```

Let's initialize the model and define the parameters:

```
model = MoiraiForecast(
    module=MoiraiModule.from_pretrained(f"Salesforce/
    moirai-1.1-R-small"),
    prediction_length=24,
    context_length=120,
    patch_size=32,
  num_samples=100,
    target_dim=1,
    feat_dynamic_real_dim=0,
    past_feat_dynamic_real_dim=0,
)
```

Let's compute the past target by passing batch, time, and variate:

```
past_target = rearrange(
    torch.as_tensor(inp["target"], dtype=torch.float32),
    "t -> 1 t 1"
)

past_observed_target = torch.ones_like(past_target,
dtype=torch.bool)

past_is_pad = torch.zeros_like(past_target, dtype=torch.
bool).squeeze(-1)
```

Let's perform forecasting:

```
forecast = model(
    past_target=past_target,
    past_observed_target=past_observed_target,
    past_is_pad=past_is_pad,
)
```

Let's calculate the error metrics:

```
calculate_error_metrics(label["target"],np.round(np.median(fore
cast[0], axis=0), decimals=4))
```

```
        MSE : 5319.221497305669
        RMSE : 72.93299320133289
        MAPE : 0.10728392512332686
        r2 : 0.04616826793581874
        adjusted_r2 : 0.0028122801147195675
```

Let's print the values of median prediction and ground truth:

```
print(
    "median prediction:\n",
    np.round(np.median(forecast[0], axis=0), decimals=4),
)
print("ground truth:\n", label["target"])
```

```
median prediction:
 [379.8565 386.5126 390.0413 398.3502 403.2256 405.6777 406.6938 417.8278
 397.7138 394.6458 414.2654 406.3108 404.5158 421.6768 425.0043 417.1104
 443.4794 433.2636 458.7923 447.1753 460.982  429.9554 452.9716 423.3293]
ground truth:
 [360 342 406 396 420 472 548 559 463 407 362 405 417 391 419 461 472 535
 622 606 508 461 390 432]
```

Let's visualize the predictions:

```
df_test = df["target"][120:144]
df_train = df["target"][:120]
df_test = df_test.reset_index().rename(columns={"index":"ds"})
df_train = df_train.reset_index().rename(columns={"ind
ex":"ds"})
df_test['Predicted']= pd.Series(np.round(np.median(forecast[0],
axis=0), decimals=4))
df_train.set_index('ds',inplace =True)
df_test.set_index('ds',inplace =True)
plt.figure(figsize=(20, 5))
y_past = df_train["target"]
y_pred = df_test['Predicted']
y_test = df_test["target"]
plt.plot(y_past, label="Past time series values")
plt.plot(y_pred, label="Forecast")
plt.plot(y_test, label="Actual time series values")
plt.title('AirPassengers Forecast', fontsize=10)
```

```
plt.ylabel('Monthly Passengers', fontsize=10)
plt.xlabel('Timestamp [t]', fontsize=10)
#plt.tight_layout()
plt.xticks(rotation=90)
plt.legend();
```

Figure 7-2 helps us to appreciate that the air passenger count predicted by our model is close to reality.

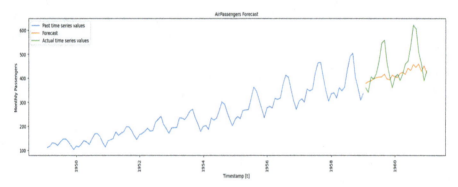

Figure 7-2. *Observed vs. forecast*

7.4 Summary

We understood how MOIRAI was developed to tackle the challenges with universal forecasting, namely, handling various frequencies of data, flexibility to support a range of variates, and producing probabilistic forecasts for multiple scenarios having datasets of different statistical and probabilistic distributions. We also saw practical implementations for univariate and multivariate scenarios.

7.5 Reference

[1].   Unified Training of Universal Time Series Forecasting Transformers by Gerald Woo et al. https://doi.org/10.48550/arXiv.2402.02592

CHAPTER 8

TimesFM: Time Series Forecasting Using Decoder-Only Foundation Model

Chapter Goal: Learn how a foundation model is designed, the challenges and approaches to solve the challenges, using decoder only design.

8 Introduction

After understanding the design, working, and developments in foundation models for time series forecasting, it is natural for us (time series forecasters) to expect models that would work out of the box. Let us understand TimesFM, one such foundation model capable of zero-shot forecasting. This model is developed based on a decoder-only transformer architecture, using input patching. The forecasting capabilities of this foundation model are comparable to supervised time series forecasting models. The datasets used for training are a combination of real and

© Banglore Vijay Kumar Vishwas and Sri Ram Macharla 2025
B. V. Vishwas and Sri Ram Macharla, *Time Series Forecasting Using Generative AI*,
https://doi.org/10.1007/979-8-8688-1276-7_8

synthetic data. The sources of data primarily include page view stats, a Wikipedia tool that provides data related to wiki page visits, and Google trends.

The decoder-only transformer consists only of the decoder stage and is best suited for autoregressive tasks. This transformer model focuses on predicting the next token in the sequence based on the tokens that were generated earlier. Please go through [2] in the "References" section for a detailed explanation of the benefits of using decoder-only models.

The challenges with designing a time series forecasting model capable of zero-shot forecasting are different when compared to the models in NLP and vision domains. There are some types of bounds while dealing with the natural language. Any language has rules, like grammar and limitation of alphabets – 26 alphabets in English, 56 in Telugu, and 49 in Kannada. Any image can be described with a finite number of pixels. Each pixel can be broken down into three components of RGB colors ranging from 0 to 255 (maximum brightness). However, time, as we all know, has no beginning or end.

8.1 Technical Overview of TimesFM

Let us understand the problem TimesFM is trying to solve. The intention behind creating a new time series foundation model, TimesFM, was to develop a zero-shot forecasting model. This general-purpose time series forecasting model, with zero-shot forecasting capabilities, takes past values of time series data as context to come up with forecasts for the future. The challenge here is that during training we cannot have covariates specific to a dataset, since the intention is to come up with a one-shot general-purpose forecaster. The problem that TimesFM is trying to solve is to learn how to generalize forecasting based on historical values, irrespective of time series properties like granularity, trend, and seasonality. This model uses the MAE (mean absolute error) metric to measure prediction results.

The concepts leveraged while coming up with this model are (**i**)
decoder-only model, (**ii**) patching, (**iii**) length of output patches, and (**iv**)
patch masking. Let us understand these concepts in more detail.

In the decoder-only design, the model is trained to predict the next
path based on earlier patches. Training is done in parallel, spanning
the entire context window. This helps the model to generalize and thus
perform time series forecasting based on different input patches learned.

A patch-based approach, which was discussed in earlier chapters,
is used by TimesFM too. Time series data is split into patches and used
during the training of this model. Patch-based learning helps improve
performance and inference speed as the number of tokens fed to the
decoder is reduced.

The length of output patches being longer helps in scenarios
demanding forecasting into longer time periods like looking into forecasts
far into the future, years, and decades ahead. A general-purpose forecaster
has to cater to longer-term predictions too. While predicting the full
horizon (forecasting far ahead into the future) yields better results, in
the case of zero-shot forecasting the forecasting time step details are not
known up front. In TimesFM, the model uses output patches that are
longer than input patches.

Patch masking helps overcome the problem of overfitting. Some
patches are randomly hidden during training. If this is not done, then
models tend to learn based on the input patch length. The forecasting
accuracy can be seen only in instances where context length is a multiple
of input patch length. TimesFM uses a random masking technique during
training. This helps the model to learn all possible context lengths during
training, ranging from 1 all the way to the maximum context length.

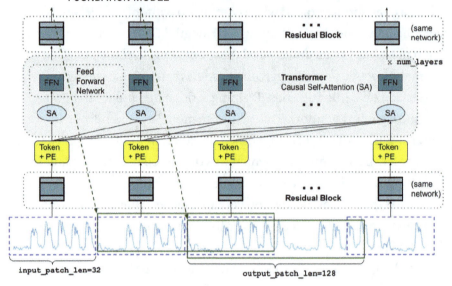

Figure 8-1. *Overall architecture of TimesFM during training [1]*

[In Figure 8-1, we can see time series data fed as input, split into patches of size input_patch_len. The residual block converts patches into vectors. A combination of positional encodings and vectors are input to the stacked transformer layers. Remember that vectors are numerical representations of tokens in the form of n-dimensional arrays. Tokens are converted to vector embeddings, thus helping models to process and compute operations on tokens. The resultant tokens are then fed to the residual block. The residual block transforms the tokens to an output patch of length output_patch_len. This output represents the forecast for the time window immediately following the last input patch processed by the model.]

Let us understand the architectural blocks and their actions in more detail.

The **input layer** processes the time series data to tokens. These tokens are converted to patches that are fed to a residual block. The residual block transforms these patches into a vector whose length is based on model dimensions. Vectors are numerical representations of tokens in high-dimensional space. The residual block consists of a multilayer perceptron with one hidden layer and a skip connection. The skip connection helps in adding the output to the next layer directly.

The **stacked transformer** architecture is used where parameters in the model are in transformer layers stacked on top of each other. The layers leverage multi-head self-attention proceeded by a feedforward network. Here, the transformer architecture uses the mechanism of causal attention. This ensures the model only considers the tokens that have arrived before the current token in the sequence. This causal attention technique is particularly helpful in scenarios where only past information (and not future tokens) should be used to predict the next token.

The final task of prediction is taken care of by the **output layers**, where the output tokens are mapped into predictions. The output tokens represent the model's understanding of data at various points in time; however, they cannot be directly interpreted for forecasting. The output tokens are mapped using a residual block. The residual block transforms the tokens to an output patch of length output_patch_len. This output represents the forecast for the time window immediately following the last input patch processed by the model. Training using the decoder-only model enables each output token to be capable of predicting the portion of the time series that follows its corresponding input patch.

8.2 TimesFM in Action

Having established a high-level theoretical foundation of TimesFM, we shall now translate abstract concepts into practical code implementation.

8.2.1 Univariate Use Case

Let us consider a univariate scenario first.

Import required modules:

```
import numpy as np
import pandas as pd
import timesfm
import matplotlib.pyplot as plt
from sklearn.metrics import mean_squared_error, mean_absolute_
percentage_error, r2_score

import os
os.environ['XLA_PYTHON_CLIENT_PREALLOCATE'] = 'false'
os.environ['JAX_PMAP_USE_TENSORSTORE'] = 'false'
```

Let's load the AirPassengersDataset CSV using pandas, a dataset that
contains 12 years of monthly air passenger data:

```
Y_df = pd.read_csv('AirPassengersDataset.csv')
Y_df['ds'] = pd.to_datetime(Y_df['ds'])
Y_df = Y_df.reset_index(drop=True)
Y_df.head()
```

	unique_id	ds	y
0	1	1949-01-31	112
1	1	1949-02-28	118
2	1	1949-03-31	132
3	1	1949-04-30	129
4	1	1949-05-31	121

Split the data into train and test.

```
Y_train_df = Y_df[Y_df.ds<='1959-12-31']
Y_test_df = Y_df[Y_df.ds>'1959-12-31']
```

Configure the model to use CPU:

```
timesfm_backend = "cpu"

from jax._src import config
config.update(
    "jax_platforms", {"cpu": "cpu", "gpu": "cuda", "tpu": ""}
    [timesfm_backend]
)
```

Initialize the TimesFM model and define the parameters:

```
tfm = timesfm.TimesFm(
    context_len=128,
    horizon_len=12,
    input_patch_len=32,
    output_patch_len=128,
    num_layers=20,
    model_dims=1280,
    backend=timesfm_backend,
)
```

Let's understand the variables used:

context_len is the length of the context window for the model.

horizon_len is the length of the forecasting horizon.

input_patch_len is the length of input patches.

output_patch_len is the length of output patches.

inputs is the dataframe containing the training time series data.

freq is the frequency of the time series data (e.g., monthly).

value_name is the name of the column with the values to be forecasted.

num_jobs is the number of parallel jobs to use for forecasting (–1 uses all available cores).

Load the pre-trained model from the checkpoint:

```
tfm.load_from_checkpoint(repo_id="google/timesfm-1.0-200m")
```

Generate forecasts using the TimesFM model on the given DataFrame:

```
timesfm_forecast = tfm.forecast_on_df(
    inputs=Y_train_df,
    freq="MS",
    value_name="y",
    num_jobs=-1,
)
timesfm_forecast = timesfm_forecast[["ds","timesfm"]]

timesfm_forecast.head()
```

	ds	timesfm
0	1960-01-01	410.117157
1	1960-02-01	381.156982
2	1960-03-01	441.656921
3	1960-04-01	432.986389
4	1960-05-01	450.837585

Evaluate how well the model works on test data:

```
calculate_error_metrics(Y_test_df[['y']],timesfm_
forecast['timesfm'])
```

```
MSE : 770.0660279974496
RMSE : 27.75006356744881
MAPE : 0.04984703514028369
r2 : 0.8609856684706817
adjusted_r2 : 0.8470842353177499
```

Let's visualize the predictions:

```
Y_train_df.set_index('ds',inplace =True)
timesfm_forecast.set_index('ds',inplace =True)
Y_test_df.set_index('ds',inplace =True)
plt.figure(figsize=(20, 5))
y_past = Y_train_df["y"][-50:]
y_pred = timesfm_forecast['timesfm']
y_test = Y_test_df["y"]
plt.plot(y_past, label="Past time series values")
plt.plot(timesfm_forecast, label="Forecast")
plt.plot(y_test, label="Actual time series values")
plt.title('AirPassengers Forecast', fontsize=10)
plt.ylabel('Monthly Passengers', fontsize=10)
plt.xlabel('Timestamp [t]', fontsize=10)
plt.tight_layout()
plt.xticks(rotation=90)
plt.legend();
```

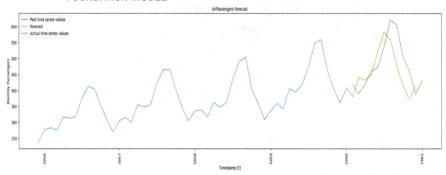

Figure 8-2. *Observed vs. forecast*

Figure 8-2 helps us to appreciate that the air passenger count predicted by our model is close to reality.

8.2.2 Multivariate Use Case

Let's now consider a multivariate scenario.

Import the necessary libraries:

```
import numpy as np
import pandas as pd
import timesfm
import matplotlib.pyplot as plt
from sklearn.metrics import mean_squared_error, mean_absolute_
percentage_error, r2_score
from collections import defaultdict

import os
os.environ['XLA_PYTHON_CLIENT_PREALLOCATE'] = 'false'
os.environ['JAX_PMAP_USE_TENSORSTORE'] = 'false'
```

Load the dataset for electricity price forecasting and create a
dataframe:

```
df = pd.read_csv('EPF_FR_BE.csv')
df[df['unique_id'] =='FR']
```

	unique_id	ds	y	gen_forecast	system_load	week_day
0	FR	2015-01-01 00:00:00	53.48	76905.0	74812.0	3
1	FR	2015-01-01 01:00:00	51.93	75492.0	71469.0	3
2	FR	2015-01-01 02:00:00	48.76	74394.0	69642.0	3
3	FR	2015-01-01 03:00:00	42.27	72639.0	66704.0	3
4	FR	2015-01-01 04:00:00	38.41	69347.0	65051.0	3
...	...	...	...	...	...	...
16075	FR	2016-10-31 19:00:00	63.89	55041.0	59537.0	0
16076	FR	2016-10-31 20:00:00	61.99	53535.0	53823.0	0
16077	FR	2016-10-31 21:00:00	52.70	49893.0	50622.0	0
16078	FR	2016-10-31 22:00:00	50.60	49037.0	49194.0	0
16079	FR	2016-10-31 23:00:00	45.47	49094.0	53441.0	0

16080 rows × 6 columns

Create a data pipeline:

```
def get_batched_data_fn(
    batch_size: int = 128,
    context_len: int = 120,
    horizon_len: int = 24,
):
  examples = defaultdict(list)

  num_examples = 0
  for country in ("FR", "BE"):
    sub_df = df[df["unique_id"] == country]
```

```python
    for start in range(0, len(sub_df) - (context_len +
    horizon_len), horizon_len):
        num_examples += 1
        examples["country"].append(country)
        examples["inputs"].append(sub_df["y"][start:(context_end
        := start + context_len)].tolist())
        examples["gen_forecast"].append(sub_df["gen_forecast"]
        [start:context_end + horizon_len].tolist())
        examples["week_day"].append(sub_df["week_day"]
        [start:context_end + horizon_len].tolist())
        examples["outputs"].append(sub_df["y"][context_
        end:(context_end + horizon_len)].tolist())

  def data_fn():
    for i in range(1 + (num_examples - 1) // batch_size):
      yield {k: v[(i * batch_size) : ((i + 1) * batch_size)]
      for k, v in examples.items()}

  return data_fn
```

Configure the model to use CPU:

```python
timesfm_backend = "cpu"
```

```python
from jax._src import config
config.update(
    "jax_platforms", {"cpu": "cpu", "gpu": "cuda", "tpu": ""}
    [timesfm_backend]
)
```

Create the model using TimesFm and pass the values:

```python
model = timesfm.TimesFm(
    context_len=512,
    horizon_len=128,
```

```
    input_patch_len=32,
    output_patch_len=128,
    num_layers=20,
    model_dims=1280,
    backend=timesfm_backend,
)
model.load_from_checkpoint(repo_id="google/timesfm-1.0-200m")
```

Let us forecast the required horizon:

```
batch_size = 128
context_len = 120
horizon_len = 24
input_data = get_batched_data_fn(batch_size = 128)
metrics = defaultdict(list)
import time

for i, example in enumerate(input_data()):
  raw_forecast, _ = model.forecast(
      inputs=example["inputs"], freq=[0] *
      len(example["inputs"])
  )
  start_time = time.time()

  cov_forecast, ols_forecast = model.forecast_with_covariates(
      inputs=example["inputs"],
      dynamic_numerical_covariates={
          "gen_forecast": example["gen_forecast"],
      },
      dynamic_categorical_covariates={
          "week_day": example["week_day"],
      },
      static_numerical_covariates={},
```

```
    static_categorical_covariates={
        "country": example["country"]
    },
    freq=[0] * len(example["inputs"]),
    xreg_mode="xreg + timesfm",
    ridge=0.0,
    force_on_cpu=False,
    normalize_xreg_target_per_input=True,
)
print(
    f"\rFinished batch {i} linear in {time.time() - start_
    time} seconds",
    end="",
)
```

Let's see the results without covariates:

```
print("Without covariates: \n")
calculate_error_metrics(raw_forecast[:, :horizon_len],
example["outputs"])
```

```
        Without covariates:

        MSE : 404.6088033992598
        RMSE : 20.114890091652498
        MAPE : 0.18969476358665338
        r2 : -0.04918699445529159
        adjusted_r2 : -0.07199540737823273
```

The results with covariates:

```
print('With covariates: \n')
calculate_error_metrics(cov_forecast, example["outputs"])
```

```
MSE : 259.9858694585536
RMSE : 16.124077321154026
MAPE : 0.15689309439567115
r2 : 0.16291603857224796
adjusted_r2 : 0.14471856114990556
```

Results with ordinary least square:

```
print('ols forecast: \n')
calculate_error_metrics(ols_forecast, example["outputs"])
```

```
ols forecast:

MSE : 3222.492537935656
RMSE : 56.767002192608835
MAPE : 477.1218213465076
r2 : -9241.742381331851
adjusted_r2 : -9442.671563534717
```

8.3 Summary

We understood how and why TimesFM was developed for general-purpose
zero-shot forecasting. The use of decoder architecture and patching was
discussed. Finally, we implemented use cases using the TimesFM model
for both univariate and multivariate scenarios.

8.4 Conclusion

Your interest in this journey to explore advancements in time series and
staying with us until the end of this book is greatly appreciated.

We have learned the evolution of LLMs starting from the basic perceptron to the latest foundation models.

From our experience working in different domains, diverse datasets, and techniques, we can say it is definitely worth trying these models alongside traditional and neural network–based models. As a best practice, start with traditional models and then move to the advanced ones.

We encountered multiple scenarios where traditional techniques outperformed foundation models. A Kannada (Indian language) proverb says "***Gubbi mele Brahmastra***," meaning using a huge weapon like Brahmastra on a tiny sparrow. So let us ensure that we use the right techniques as per use case and datasets.

We discussed techniques that help in repurposing existing foundation models. We hope with the theory and implementation knowledge gained so far, you will apply and appreciate these models in real-time scenarios.

The future of AI seems very promising, and we see many new foundation models popping up, such as Tiny Time Mixtures, MOMENT, MambaTS, Lag-Llama, and timer-base-84m.

Happy Learning!!!

8.5 Reference

[1].   A Decoder-Only Foundation Model for Time-Series Forecasting by Abhimanyu Das et al. https://doi.org/10.48550/arXiv.2310.10688

Index

A

Any-variate
    attention, 187
ARIMA, *see* Autoregressive
    Integrated Moving
    Average (ARIMA)
Attention function, 88
Autoformer model, 110
AutoGluon, 32
Autoregression (AR), 64
    DeepAR, 64–73
    probabilistic forecasting
        features, 64
    RNNs, 68
Autoregressive Integrated Moving
    Average (ARIMA), 1, 5
Autoregressive
    models, 32

B

Bidirectional temporal
    convolutional network
    (BiTCN), 44
BiTCN, *see* Bidirectional temporal
    convolutional
    network (BiTCN)

C

Chronos, 156
Convolutional neural
    networks (CNNs), 7
    BiTCN, 44–50
    computer vision/speech
        processing, 27
    TCN, 36–40, 42–44
    WaveNet architecture, 28–36
CNNs, *see* Convolutional neural
    networks (CNNs)

D

Decoder-only foundation
    model, TimesFM
    architecture, 198, 199
    challenges, 196
    multivariate use case,
        204–207, 209
    patch-based approach, 197
    patch masking, 197
    practical code implementation,
        199–201, 203, 204
    zero-shot forecasting, 195, 196
Deep learning models, 32
Dot product, 89

© Banglore Vijay Kumar Vishwas and Sri Ram Macharla 2025
B. V. Vishwas and Sri Ram Macharla, *Time Series Forecasting Using Generative AI*,
https://doi.org/10.1007/979-8-8688-1276-7

E

Encoder-decoder architectures, 84

F

Feedforward network, 105
Feedforward neural
        network (FFN), 91
FFN, *see* Feedforward neural
        network (FFN)
Foundation models, 17

G, H

GARCH, *see* Generalized
        Autoregressive Conditional
        Heteroskedasticity
        (GARCH) models
Gated Recurrent Unit (GRU), 7
Gemini AI models, 12
Generalized Autoregressive
        Conditional
        Heteroskedasticity
        (GARCH) models, 7, 169
Generative AI
    evolution from AI, 9–12
    generate novel content, 9
    techniques, 8
    time series, 13, 14
GRU, *see* Gated Recurrent
        Unit (GRU)

I, J, K

Informers, 95

L

Large language model (LLM), 15
LLM, *see* Large language
        model (LLM)
LLM-based forecasting
        models, 156
Long sequence time series
        forecasting (LSTF), 94
Long short-term memory
        (LSTM), 7, 57
LSTF, *see* Long sequence time
        series forecasting (LSTF)
LTSF-Linear, 109
LSTM, *see* Long short-term
        memory (LSTM)

M

Masked EncOder-based UnIveRsAl
        Time Series Forecasting
        Transformer (MOIRAI)
    any-variate attention, 187
    architecture, 187
    distribution, 188
    practical code
        implementation, 188–194
    traditional time series, 184

universal forecasting model,
challenges, 184–186
zero-shot time series, 184
Mean absolute error (MAE), 196
MLP, *see* Multilayer
perceptron (MLP)
MOIRAI, *see* Masked EncOder-
based UnIveRsAl Time
Series Forecasting
Transformer (MOIRAI)
Multi-head attention, 91
Multilayer perceptron (MLP), 21
Multivariate time series
analysis, 6

N, O

Natural language
processing (NLP), 51
NBEATS, *see* Neural basis
expansion analysis for time
series (NBEATS)
Neural basis expansion analysis for
time series (NBEATS)
advantages, 75
MLPs, 74
practical code
implementation, 77–80
residual stacking principle, 76
stacks, 75
Neural networks, *see* Perceptron
NLinear, 118
NLP, *see* Natural language
processing (NLP)

P, Q

Patch-based approach, 197
PatchTST, 122, 130
Perceptron
AR, 64
CNNs, 27
example, 18
formula, 19
foundation models, 17
sequential data
LSTM, 57, 58, 60–63
RNN, 51–57
training phase, 19, 20, 22,
23, 25–27
Phase-Structure Grammar, 10
Positional encoding, 86
Pre-trained probabilistic time
series model
Chronos, 156
large_ensemble use
case, 163–167
tiny use case, 159–163
inference, 159
LLMs, 155
time series tokenization,
157, 158
training, 158
ProbSparse self-attention
mechanism, 96

R

Recurrent language models, 84

Recurrent neural networks (RNNs),
        7, 51, 66
RNNs, *see* Recurrent neural
        networks (RNNs)

S

SARIMA, *see* Seasonal
        Autoregressive Integrated
        Moving Average (SARIMA)
SARIMAX, *see* Seasonal
        Autoregressive Integrated
        Moving Average with
        Exogenous Regressors
        (SARIMAX)
Seasonal Autoregressive Integrated
        Moving Average
        (SARIMA), 5
Sequence learning
        architectures, 84
Supervised time series forecasting
        models, 195
SyntheMol AI model, 12

T

TCN, *see* Temporal convolutional
        network (TCN)
Temporal convolutional
        network (TCN), 36
TimeGPT, 131
    API key configuration process,
        setting up, 173
    characteristics, 171

deep learning models, 170
encoder-decoder
        architecture, 171
    forecasting model, 172
    foundation model, 169
    machine learning tools, 170
    multivariate dataset, 178–182
    single-series/multiple-series
        forecasting, 171
    transformer-based models, 170
    univariate dataset, 175–178
    zero-shot inference, 173
Time-LLM
    fine-tuning *vs.*
        reprogramming, 132
        foundation model, 132
        framework, 133
        multivariate use case,
            144–148, 150–153
        univariate problem, 139–144
        working, 134–138
    foundation model, 131
    framework, 132
    NLP applications, 131
Time series analysis
    characteristics, 3
    fields, 2
    forecasting methods, 4–7, 16
    IT industry, 1
    neural networks, 17
    real-world examples, 2
    simple ARIMA, 1
Transformer-based methods, 44
Transformers, 60

architecture, 85

components, 86, 87,
    89–91, 93, 94

DLinear
    architecture, 111–113
    autoformer, 110
    FEDformer, 110
    linear models, 109
    practical code
        implementation, 114–117

encoder and decoder, 84

inverted
    attention
        mechanisms, 103–105
    linear forecasting
        models, 102
    practical code
        implementation, 105–108

natural language processing, 83

NLinear, 118–120, 122

PatchTST
    channel independence, 122
    multivariate time series
        data, 123
    practical code
        implementation, 125–128
    representation learning, 125
    subseries-level patches, 122
    transformer encoder, 124

vanilla, 94

U

Univariate time series analysis, 4

V

Vanilla transformer
    drawbacks, 94
    features, 95
    informers, 95, 97
    NLP, 129
    practical code implementation,
        98–102
    RNN models, 94

VAR, *see* Vector
    Autoregression (VAR)

VARMA, *see* Vector Autoregressive
    Moving Average (VARMA)

VARMAX, *see* Vector
    Autoregression Moving
    Average with Exogenous
    Regressors (VARMAX)

VECM, *see* Vector Error Correction
    Model (VECM)

Vector Autoregression (VAR), 6

Vector Autoregressive Moving
    Average (VARMA), 6

Vector Error Correction
    Model (VECM), 7

W, X, Y

WaveNet, 28, 32

Z

Zero-shot forecasting,
    155, 195

GPSR Compliance

The European Union's (EU) General Product Safety Regulation (GPSR) is a set of rules that requires consumer products to be safe and our obligations to ensure this.

If you have any concerns about our products, you can contact us on

ProductSafety@springernature.com

In case Publisher is established outside the EU, the EU authorized representative is:

Springer Nature Customer Service Center GmbH
Europaplatz 3
69115 Heidelberg, Germany

www.ingramcontent.com/pod-product-compliance
Lightning Source LLC
LaVergne TN
LVHW051638050326
832903LV00022B/806